Springer Theses

Recognizing Outstanding Ph.D. Research

Aims and Scope

The series "Springer Theses" brings together a selection of the very best Ph.D. theses from around the world and across the physical sciences. Nominated and endorsed by two recognized specialists, each published volume has been selected for its scientific excellence and the high impact of its contents for the pertinent field of research. For greater accessibility to non-specialists, the published versions include an extended introduction, as well as a foreword by the student's supervisor explaining the special relevance of the work for the field. As a whole, the series will provide a valuable resource both for newcomers to the research fields described, and for other scientists seeking detailed background information on special questions. Finally, it provides an accredited documentation of the valuable contributions made by today's younger generation of scientists.

Theses are accepted into the series by invited nomination only and must fulfill all of the following criteria

- They must be written in good English.
- The topic should fall within the confines of Chemistry, Physics, Earth Sciences, Engineering and related interdisciplinary fields such as Materials, Nanoscience, Chemical Engineering, Complex Systems and Biophysics.
- The work reported in the thesis must represent a significant scientific advance.
- If the thesis includes previously published material, permission to reproduce this must be gained from the respective copyright holder.
- They must have been examined and passed during the 12 months prior to nomination.
- Each thesis should include a foreword by the supervisor outlining the significance of its content.
- The theses should have a clearly defined structure including an introduction accessible to scientists not expert in that particular field.

More information about this series at http://www.springer.com/series/8790

Bernhard Rauer

Non-Equilibrium Dynamics Beyond Dephasing

Recurrences and Loss Induced Cooling in One-dimensional Bose Gases

Doctoral Thesis accepted by
the Atominstitut, TU Vienna, Vienna, Austria

Springer

Author
Dr. Bernhard Rauer
Laboratoire Kastler Brossel
École Normale Supérieure
Paris, France

Supervisor
Prof. Hannes-Jörg Schmiedmayer
Atominstitut
TU Vienna
Vienna, Austria

ISSN 2190-5053 ISSN 2190-5061 (electronic)
Springer Theses
ISBN 978-3-030-18235-9 ISBN 978-3-030-18236-6 (eBook)
https://doi.org/10.1007/978-3-030-18236-6

This Springer imprint is published by the registered company Springer Nature Switzerland AG.
The registered company address is: Gewerbestrasse 11, 6330 Cham, Switzerland

Supervisor's Foreword

The question of whether an isolated quantum many-body system relaxes is a central problem at the interface between microscopic quantum dynamics and statistical physics. It is of high relevance in many diverse fields, ranging from decoherence in quantum information and metrology to the complex dynamics in high-energy physics and cosmology. Moreover, relaxation processes are intimately related to the question if and how the classical world at the macro-scale can emerge from the unitary quantum evolution at the micro-scale.

The thesis of Bernhard Rauer investigates a series of questions that are highly contested in this context: How does an isolated quantum many-body system relax? To what extent and through which processes is the memory of an initial state erased during time evolution? How does a classical ensemble description in the sense of statistical physics emerge from the underlying quantum evolution when an actual environment is absent?

Over the last years, ultracold gases have emerged as an ideal model system to study these questions. They are nearly perfectly isolated and a large set of tools is available to manipulate and probe their many-body physics in a controlled setting. One-dimensional (1d) gases are especially interesting in that context. They exhibit very rich dynamics and can be compared to well-studied theoretical models. An especially interesting aspect is that in many 1d systems a large number of conserved quantities constrain the evolution and have a profound effect on their relaxation processes.

In his thesis, Bernhard Rauer investigates relaxation dynamics in degenerate 1d Bose gases of ^{87}Rb atoms, confined on an atom chip. The thesis studies two main topics: the first being concerned with the creation of these systems in the laboratory. In experimental realizations of 1d Bose gases, the temperature of the system needs to be brought below the minimal energy of transverse excitations to confine dynamics to the desired single dimension. However, this leads to an exponential suppression of thermalizing collisions which should force thermalization to stop. Nevertheless, evaporative cooling, which highly depends on efficient thermalization processes, works surprisingly well within the 1d regime and experiments reach temperatures far below the freeze-out of thermalizing collisions. Why these low temperatures can be reached remained a mystery for the last 10 years. Bernhard Rauer investigates the cooling process in detail and presents a semi-classical mechanism that describes the cooling as a loss-driven reduction of fluctuations under a continuous dephasing of the involved phononic excitations. This model is in good agreement with the experimental findings and finally provides a solution to the cooling conundrum. Remarkably, the contribution of atomic shot noise that is expected to become relevant at low temperatures seems to be absent in the measurements.

The second focus of the thesis lies in the out-of-equilibrium dynamics following a quench. Extending Poincaré's recurrence theorem to the quantum domain, any finite quantum system should at some point return arbitrarily close to its initial state. In large many-body systems, however, the complexity of the many-body eigenstates and their spectrum leads to exceedingly long recurrence times. Yet, Bernhard Rauer was able to observe recurrences of coherence in the post-quench relaxation dynamics of a pair of 1d Bose gases. The key to the experiment lies in the insight that one cannot observe the many-body eigenstates directly, but can only observe much simpler few-body observables. These observables can be linked to an effective quantum field theory description through the system's collective modes. Controlling the dispersion of these collective modes, in this case phonons, allows to observe recurrences in experimentally accessible quantities. Performing the post-quench relaxation experiments in a box-shaped potential results in commensurate phonon frequencies such that the observation of their rephasing becomes feasible, even for systems of thousands of particles. Both the recurrence time and its scaling with the size of the system are well described by a low-energy Luttinger liquid model. In contrast, the measured damping of the recurrence signal stems mainly from phonon-phonon scattering processes mediated by terms beyond the low-energy effective field theory description. The presented experiment sets a beautiful example of how rephasing can be used to probe many-body processes at times beyond the initial dephasing dynamics, opening up a new window to study quasi-particle interactions in such systems. Another remarkable feature of the experiment is that, even though the system apparently relaxes to a classical (generalized) Gibbs ensemble, coherence underneath this classical statistical physics cover leads to a return of the initial state at the point of the recurrence. This illustrates the intricate connection between the unitary evolution in quantum many-body systems and the description of their relaxed state in terms of statistical mechanics.

Vienna, Austria
March 2019

Prof. Dr. Hannes-Jörg Schmiedmayer

Abstract

Out-of-equilibrium dynamics in complex quantum many-body systems is a vast topic of research touching many different areas of physics. One of the most versatile experimental platforms to investigate these effects are ultracold atoms, due to their flexibility and easy isolation from the environment. In this thesis, we investigate non-equilibrium dynamics of one-dimensional (1d) Bose gases realized with ultracold ^{87}Rb atoms on an atom chip. Focusing on phenomena emerging on timescales beyond the typical dephasing times of excitations, we report on the observation of recurrences and the finding of a novel cooling mechanism.

A recurrence, the dynamic return of a system to its initial state, can generally not be observed in large systems as the complexity of their excitation spectra shifts its appearance to prohibitively long times. Yet, by realizing a commensurate spectrum in a pair of near-homogeneous 1d Bose gases, recurrences in their low-energy dynamics can be observed on experimentally accessible timescales. We demonstrate this by initializing two gases in a phase coherent state by coupling them through a tunneling barrier before suddenly ramping the coupling to zero. The subsequent dynamics is monitored by matter-wave interferometry, providing access to the relative phase field between the gases. After an initial dephasing dynamics, we observe multiple recurrences of the coherent initial state due to a rephasing of the underlying excitations. Additionally, analyzing the damping of these recurrences, we detect otherwise elusive scattering effects between excitations.

Furthermore, we investigate the dynamics of a 1d Bose gas under a continuous loss of particles. With thermalization strongly inhibited in these systems, standard evaporative cooling is rendered ineffective; yet, we still observe a substantial cooling effect. This cooling is driven by a novel mechanism that relies neither on an energy selective extraction of particles nor on efficient thermalization channels. Instead, it proceeds through a loss-induced reduction of density fluctuations and a continuous dephasing of the involved excitations. For experiments with 1d Bose gases, this mechanism fills an important gap in the understanding of the state preparation and the limits of cooling.

Acknowledgements

I would like to express my gratitude to everybody involved in making this thesis possible, first and foremost the members of the KRb team. Cold atoms experiments can be complex beasts and are best tamed in groups. Without their countless hours working on the machine, the sharing of codes and the long discussions, none of the results presented here would be possible. Also, I want to thank all current and former members of the Atomchip group that I had the pleasure to work along with, for numerous borrowed equipments, stimulating discussions and for creating the most friendly work environment one can wish for. The infamous coffee corner discussions and long evenings will be missed.

In particular, I would like to thank the following people:

- My supervisor Hannes-Jörg Schmiedmayer, for giving me the opportunity to work in his group and for sharing his fascination for physics.
- Our former team members Micheal Gring, Max Kuhnert, Tim Langen, Remi Geiger, and David Adu Smith from whom I learned a lot and who left us a well-oiled machine.
- Thomas Schweigler, my main co-worker on the setup, for his razor-sharp mind always asking the right questions, and for the careful reading of most of this manuscript. Also, for organizing many alpine adventures.
- Sebastian Erne, Pjotrs Grišins, and Igor Mazets for their theory support, shared codes, and numerous discussions that helped to elucidate the investigated phenomena.
- The next generation of KRb team members taking over the experiment: Federica Cataldini, Amin Tajik, João Sabino, and SiCong Ji, who will hopefully realize many interesting projects in the coming years. Particularly, I would like to thank Amin for putting all that work into the DMD setup.
- Jacqueline Bloch for giving me the opportunity to visit her group and learn about the fascinating world of exciton-polaritons.
- Hanns-Christoph Nägerl and Iacopo Carusotto for kindly agreeing to review my thesis.

- The Austrian and European taxpayer for supporting basic research and funding the work presented here.

Zu guter Letzt möchte ich meinen Freunden und meiner Familie danken, speziell meinen Eltern für ihre kontinuierliche Unterstützung, die mich zu diesem Punkt gebracht hat.

Der größte Dank aber gebührt meiner Partnerin Jaqueline für ihre fortwährende Unterstützung und Motivation, trotz der vielen Abende, die sie ohne mich verbringen musste damit dieses Ding hier fertig wird.

Contents

Chapter 1
Preamble

The advent of ultracold atoms, realizing a multitude of condensed matter models in a well controlled and tunable fashion, marked a new era in the experimental study of interacting many-body systems [1]. Their near perfect isolation from the environment, the development of versatile probing schemes and the flexibility in geometry and interactions they exhibit make cold atoms an ideal platform for analog simulations of complex quantum many-body physics [2]. In particular, otherwise elusive out-of-equilibrium phenomena are well accessible in these systems and generated a large interest over the past years [3–5]. From the spread of correlations [6, 7] over transport properties [8, 9], the dynamics of phase transitions [10–12] to periodically driven systems [13], ensembles of cold atoms are a resourceful tool in the study of physics beyond equilibrium states.

In that context, this thesis studies two effects that emerge on timescales beyond the typical dephasing time, utilizing one-dimensional (1d) Bose gases on an atom chip as its experimental platform [14]. The first of these effects is linked to the interesting question of relaxation dynamics in isolated quantum many-body systems [15]. Following a unitary microscopic evolution it is a priori not clear how a closed system can lose memory of its initial configuration and arrive at an equilibrium or thermal state. Yet, we know quantum statistical mechanics to be a successful theory. For a large class of systems a solution to this conundrum is provided by the eigenstate thermalization hypothesis, shifting the the focus from the time evolution to the properties of the involved eigenstates, which already contain the thermal properties [16–19]. For integrable systems on the other hand, of which the 1d Bose gas is a prime example, this argumentation fails and thermalization is shown to be suppressed [20, 21]. Nevertheless, also in these systems, where a large number of conserved quantities restricts the dynamically available phase space, an equilibrium can be reached in the form of a maximum entropy state [22, 23].

Strictly speaking however, any closed system, integrable or not, will only equilibrate on average or over large intervals of time. The memory of its initial state cannot be truly lost and will at some point lead to a recurrence, i.e. to a dynamic return of

B. Rauer, *Non-Equilibrium Dynamics Beyond Dephasing*,
Springer Theses, https://doi.org/10.1007/978-3-030-18236-6_1

the initial configuration [24–26]. For all but the simplest systems, the timescale of such a return is prohibitively long, generally preventing its observation. However, for a large class of many-body systems the essential dynamical features beyond the microscopic dynamics can be described by effective field theories, exhibiting a much simple structure. This simplicity enables us to demonstrate recurrences within the low-energy dynamics of a system of thousands of particles [27]. Apart from their fundamental interest, these recurrences pose a sensitive probe into the systems coherent dynamics on timescales beyond the initial relaxation. We show how their decay can be used to detect otherwise elusive scattering effects between phononic excitations and thereby test the validity regime of our low-energy description.

The second study discussed in this thesis is motivated by questions arising for integrable systems regarding the experimental state preparation. If the final preparation steps are performed after the system reached its integrable configuration, can one ever prepare the system in a thermal state? We approach this problem for the preparation of a 1d Bose gas through evaporative cooling. Experimentally, cooling is observed for gases deep in the 1d regime, creating states well described by thermal distributions. The suppression of thermalizing collisions in these systems, however, should render standard cooling by evaporation ineffective. An explanation for these phenomena is found in a novel cooling mechanism described in Refs. [28, 29]. It relies solely on a loss-induced reduction of fluctuations paired with a continuous dephasing of the involved excitations. Each eigenmode of the system is cooled independently, thereby preserving an initialy thermal distribution. This mechanism fills an important gap in the understanding of the state preparation and the limits of cooling in experiments with 1d Bose gases and invites to think along new lines when devising cooling schemes for degenerate gases.

Outline

This thesis is structured as follows: Chap. 2 gives a brief introduction to the physics of 1d Bose gases, providing the necessary tools to describe and interpret the experimental findings. The discussion focuses on low-energy effective models and the perturbations arising from the gases transverse confinement. Following that, Chap. 3 first discusses basic experimental techniques and introduces the experimental setup. Both of these parts are kept short as extensive literature and a comprehensive body of theses exist on these topics. A final more detailed section is dedicated to probing, describing the different techniques developed to extract information on the system from absorption images of its density distributions in expansion.

Chapter 4 discusses the cooling of 1d Bose gases by a uniform loss of particals. First, the experimental observations that motivated this study are presented. They show that evaporative cooling still works surprisingly well within the 1d regime, where the suppression of thermalizing collisions should render it ineffective. In the following, we investigate the outcoupling process and conclude that in our 1d configuration the extraction of particles happens nearly homogeneously, adding to the mystery of the cooling process. We then present a novel mechanism that describes the cooling as a loss-driven reduction of fluctuations under a continuous dephasing of the involved excitations. In the limit of classical fields, this model explains the experimental findings well. Though, the contribution of atomic shot noise that is expected to become relevant at low temperatures seems absent in the measurements. The discussion of this conundrum forms the final part of the chapter.

Chapter 5 presents the observation of recurrences in the post-quench dynamics of a pair of 1d Bose gases. After a brief discussion of the conditions of mode rephasing and the optimal measurement protocols to favor its observation, the experimental findings are presented at length. It is shown that both the recurrence time and its scaling with the size of the system are well described by a low-energy Luttinger liquid model. In contrast, the measured recurrence damping is concluded to stem mainly from phonon-phonon scattering processes mediated by terms beyond the low-energy description. This is inferred from comparisons with numerical simulations, setting a striking example how rephasing effects can be used as sensitive probes for many-body processes at times beyond the initial dephasing dynamics.

Lastly, Chap. 6 provides an outlook on ongoing and future research, with a special focus on projects that are a continuation of the work presented. In particular, a new light shaping setup is discussed. It is designed to achieve arbitrary control over the atoms 1d confinement and hold the possibility of accessing multiple interesting research topics, from quantum thermal machines [30] to analogue black holes [31, 32].

References

1. Bloch I, Dalibard J, Zwerger W (2008) Many-body physics with ultracold gases. Rev Mod Phys 80:885–964
2. Bloch I, Dalibard J, Nascimbène S (2012) Quantum simulations with ultracold quantum gases. Nat Phys 8:267–276
3. Polkovnikov A, Sengupta K, Silva A, Vengalattore M (2011) Colloquium: nonequilibrium dynamics of closed interacting quantum systems. Rev Mod Phys 83:863–883
4. Eisert J, Friesdorf M, Gogolin C (2015) Quantum many-body systems out of equilibrium. Nat Phys 11:124–130
5. Langen T, Geiger R, Schmiedmayer J (2015) Ultracold atoms out of equilibrium. Annu Rev Condens Matter Phys 6:201–217
6. Cheneau M et al (2012) Light-cone-like spreading of correlations in a quantum many-body system. Nature 481:484–7
7. Langen T, Geiger R, Kuhnert M, Rauer B, Schmiedmayer J (2013) Local emergence of thermal correlations in an isolated quantum many-body system. Nat Phys 9:640–643
8. Ronzheimer JP et al (2013) Expansion dynamics of interacting bosons in homogeneous lattices in one and two dimensions. Phys Rev Lett 110:205301
9. Krinner S, Esslinger T, Brantut J-P (2017) Two-terminal transport measurements with cold atoms. J Phys Condens Matter 29:343003
10. Braun S et al (2015) Emergence of coherence and the dynamics of quantum phase transitions. Proc Natl Acad Sci 112:3641–3646
11. Navon N, Gaunt AL, Smith RP, Hadzibabic Z (2015) Critical dynamics of spontaneous symmetry breaking in a homogeneous Bose gas. Science 347:167–170

12. Erne S, Bücker R, Gasenzer T, Berges J, Schmiedmayer J (2018) Universal dynamics in an isolated one-dimensional Bose gas far from equilibrium. Nature 563:225–229
13. Jotzu G et al (2014) Experimental realization of the topological Haldane model with ultracold fermions. Nature 515:237–240
14. Reichel J, Vuletić V (eds) (2011) Atom Chips. Wiley-VCH, Weinheim, Germany
15. Gogolin C, Eisert J (2016) Equilibration, thermalisation, and the emergence of statistical mechanics in closed quantum systems. Rep Prog Phys 79:056001
16. Deutsch JM (1991) Quantum statistical mechanics in a closed system. Phys Rev A 43:2046–2049
17. Srednicki M (1994) Chaos and quantum thermalization. Phys Rev E 50:888–901
18. Rigol M, Dunjko V, Olshanii M (2008) Thermalization and its mechanism for generic isolated quantum systems. Nature 452:854–858
19. Kaufman AM et al (2016) Quantum thermalization through entanglement in an isolated many-body system. Science 353:794–800
20. Kinoshita T, Wenger T, Weiss DS (2006) A quantum Newton's cradle. Nature 440:900–903
21. Gring M et al (2012) Relaxation and prethermalization in an isolated quantum system. Science 337:1318–1322
22. Rigol M, Dunjko V, Yurovsky V, Olshanii M (2007) Relaxation in a completely integrable many-body quantum system: an ab initio study of the dynamics of the highly excited states of 1D lattice hard-core bosons. Phys Rev Lett 98:050405
23. Langen T et al (2015) Experimental observation of a generalized Gibbs ensemble. Science 348:207–211
24. Bocchieri P, Loinger A (1957) Quantum recurrence theorem. Phys Rev 107:337–338
25. Percival IC (1961) Almost periodicity and the quantal H theorem. J Math Phys 2:235
26. Hogg T, Huberman BA (1982) Recurrence phenomena in quantum dynamics. Phys Rev Lett 48:711–714
27. Rauer B et al (2018) Recurrences in an isolated quantum many-body system. Science 360:307–310
28. Rauer B et al (2016) Cooling of a one-dimensional bose gas. Phys Rev Lett 116:030402
29. Grišins P, Rauer B, Langen T, Schmiedmayer J, Mazets IE (2016) Degenerate Bose gases with uniform loss. Phys Rev A 93:033634
30. Kosloff R, Levy A (2014) Quantum heat engines and refrigerators: continuous devices. Annu Rev Phys Chem 65:365–393
31. Unruh WG (1981) Experimental black-hole evaporation? Phys Rev Lett 46:1351–1353
32. Steinhauer J (2014) Observation of self-amplifying Hawking radiation in an analogue black-hole laser. Nat Phys 10:864–869

Chapter 2
Introduction and Theoretical Basics

This chapter discusses the theoretical foundations of 1d Bose gases, geared towards the systems investigated in our setup. After a short introduction of interacting 1d Bose gases we briefly introduce the seminal Lieb-Liniger model, discussing its distinct phases before giving the quasi-condensates regime a more detailed treatment. There, our main interest lies in a low-energy perturbative expansion, extracting the relevant terms in the Hamiltonian that dominate the dynamics observable in our experimental setup. This leads us to the Luttinger liquid model whose elementary phononic excitations are discussed in detail, providing the basis for the loss cooling model introduced in Chap. 4. Another focus is laid on the description of two coupled quasi-condensates, with the treatment of their excitation spectrum laying the theoretical foundation for the discussion of the recurrence phenomena in Chap. 5. Further, perturbations to the 1d description stemming from the transverse confinement in real world implementations are discussed at length. At last, some numerical tools are introduced that allow us to efficiently simulate the systems dynamics for different levels of description.

2.1 The Interacting 1d Bose Gas

One-dimensional models of many-body systems have been a playground of mathematical and theoretical physics for decades [1]. Long thought to be purely academic, approximate real world realizations renewed the interest in these models and enabled the experimental testing of many of their predictions. These realizations range from edge states in quantum hall systems over carbon nanotubes to 1d crystal structures within 3d materials. Especially recent progress in cold atoms, creating 1d many-body systems in a tunable and clean fashion, added to this development.

Although one might expect the 1d world to be full of pathological and boring models it turns out it houses a rich and unique palatte of physics. For the case of

B. Rauer, *Non-Equilibrium Dynamics Beyond Dephasing*,
Springer Theses, https://doi.org/10.1007/978-3-030-18236-6_2

Bosons moving in a single dimension, already the ideal 1d Bose gas behaves quite different from its higher dimensional counterparts. While in 3d, a gas brought to low temperatures or high densities will eventually undergo Bose-Einstein condensation (BEC) where particles pile up in the lowest energy single particle state, in 1d the behavior is different [2]. The reason for this lies in the dimensionality dependence of the density of states. In free space, the density of states behaves as $\mathcal{D}(\varepsilon) \propto \varepsilon^{d/2-1}$, where ε is the state energy and d is the dimensionality. This means, whereas in 3d the density of state is proportional to $\sqrt{\varepsilon}$, for a 1d gas $\mathcal{D}(\varepsilon) \propto 1/\sqrt{\varepsilon}$. For $\varepsilon \to 0$ the number of available states therefore diverges, such that at lower energies there are increasingly many states found in an interval $[\varepsilon, \varepsilon + \mathrm{d}\varepsilon]$. This prevents the macroscopic occupation of a single state and leads to an inherent multimode behavior. Linked to the Mermin-Wagner theorem [3, 4] stating that for $d \leq 2$ continuous symmetries cannot be broken at finite temperatures, this is the origin of the fluctuations dominating the observations presented in this thesis.

Turning to interacting gases we observe further peculiarities of the 1d case. If two particles scatter elastically within a single dimension the combined constraints of energy and momentum conservation allow only for a swap of their momenta. Such collisions therefore do not alter the overall momentum distribution, preventing the gas from ergodically sampling the phase space shell associated to its total energy. An intuitive example from the classical world illustrating this effect is the Newton's cradle. In this popular office toy several spheres are suspended on strings allowing them to only move along one dimension. In a frictionless world, once excited, these spheres would perform the same periodic motion till eternity. A quantum analogue of this system was realized with strongly interacting Bosons showing exactly this absence of thermalization [5]. More formally, the root of this behavior lies in the integrability of the system.[1] Its dynamical evolution is constraint through the existence of multiple conserved quantities, a situation particularly interesting when studying non-equilibrium dynamics. The main results presented in this thesis are all linked to this integrable behavior.

In cold atoms, 1d Bose gases are predominantly realized in optical lattices [7–10] and tight magnetic traps created on atom chips [11–13]. In our experimental setup, we utilize the latter to trap and cool samples of ^{87}Rb, as described later in Chap. 3. At ultracold temperatures, the interactions of these bosonic atoms are limited to s-wave scattering. This means they can be modeled as point like particles whose interaction can be described by the pseudopotential $U_{\mathrm{int}}(\mathbf{r}, \mathbf{r}') = g_{\mathrm{3d}}\, \delta(\mathbf{r} - \mathbf{r}')$ [14]. Here, $\mathbf{r}$ and $\mathbf{r}'$ are the positions of the atoms and the coupling constant is given by

$$g_{\mathrm{3d}} = \frac{4\pi^2 \hbar^2 a_{\mathrm{s}}}{m}, \tag{2.1}$$

[1]Note that integrability is only well defined for the classical case. It means that a system with n degrees of freedom is restricted by n independent conserved quantities. The translation of this concept to quantum mechanics is not trivial [6]. Nevertheless, the term is generally used in the literature in reference to 1d Bose gases.

with m being the mass of the atoms and a_s being the s-wave scattering length defining the strength and sign of the interactions. For ^{87}Rb we get a positive scattering length of $a_s \simeq 5.2$ nm leading to repulsive interactions [15].[2]

In both optical lattices and atom chips, the reduction of dynamics to a single effective dimension is realized by a strong confinement of the gas in the two other dimensions. At low enough temperatures this forces the atoms to occupy the ground state of the tight transverse potential such that any motion along these dimensions is effectively frozen out. The details of this transition from 3d to 1d and the influence that the transverse degrees of freedom have on the 1d description will be discussed in Sect. 2.2. For now, let us assume that we can separately describe the 1d dynamics and that the effective contact interactions are inherited from the 3d description, such that $U_{\text{int}}(z - z') = g_{1\text{d}}\,\delta(z - z')$ with $g_{1\text{d}}$ being the effective 1d coupling constant and z being the position along the 1d axis. The resulting 1d Bose gas with contact interactions is an especially rewarding system with a rich landscape of physical regimes vastly different from higher dimensions [16–19]. It is exactly solvable, making it an ideal testbed for quantum many-body physics. In Sect. 2.1.1 we will shortly discuss its different phases, loosely following Ref. [20] which gives an intuitive overview of the topic. Later, in Sect. 2.1.2 we will concentrate on the experimentally relevant quasi-condensate regime and derive an effective model for the parameter regime accessible to our setup. Section 2.1.3 then extends this description to two coupled quasi-condensates.

2.1.1 Lieb-Liniger Model

In one dimension, Bosons interacting with a repulsive contact interaction are described by the seminal Lieb-Lininger Hamiltonian [16]

$$\hat{H} = -\frac{\hbar^2}{2m} \int \mathrm{d}z\, \hat{\psi}^\dagger(z)\, \partial_z^2\, \hat{\psi}(z) \,+\, \frac{g_{1\text{d}}}{2} \int \mathrm{d}z\, \hat{\psi}^\dagger(z)\hat{\psi}^\dagger(z)\hat{\psi}(z)\hat{\psi}(z)\,, \tag{2.2}$$

with $\hat{\psi}(z)$ being the field operator in second quantization, annihilating a particle at position z. Likewise, $\hat{\psi}^\dagger(z)$ creates a particle at position z. From the coupling constant $g_{1\text{d}}$ one can construct an intrinsic length scale of the problem, $l_g = \hbar^2/m g_{1\text{d}}$. The ratio of the mean inter-particle distance $1/n_{1\text{d}}$, with $n_{1\text{d}}$ being the linear density, and this length scale then gives the famous Lieb-Liniger parameter γ determining the interaction strength

$$\gamma = \frac{1}{n_{1\text{d}}\, l_g} = \frac{m g_{1\text{d}}}{\hbar^2 n_{1\text{d}}}\,. \tag{2.3}$$

---

[2]This particular value is obtained if both atoms are in the internal state $|F = 2, m_F = 2\rangle$, which is the case in our setup (see Sect. 3.1.1).

Note that counterintuitively interactions in this model become stronger when the density is reduced. At zero temperature, in the regime of weak interactions with $\gamma \ll 1$ the gas is in a quasi-condensed state where density fluctuations are suppressed, as in 3d condensates, but phase fluctuations prevail. This is due to the fact that no macroscopic population of the lowest energy mode is possible in 1d, as discussed above. For $\gamma \gg 1$ the system becomes strongly interacting, entering the so-called Tonks-Girardeau regime. There, particles act like impenetrable spheres and can be mapped onto spinless Fermions in the limit of $\gamma \to \infty$, their strong repulsion mimicking the Pauli exclusion principle [18].

At finite temperatures the system can still be solved analytically [19]. Similarly to l_g an intrinsic energy scale is constructed by $E_g = m g_{1d}^2 / 2\hbar^2$ through which the thermal occupation of excitation can be parametrized by

$$t = \frac{k_B T}{E_g} = \frac{2\hbar^2 k_B T}{m g_{1d}^2}\,, \tag{2.4}$$

with T being the temperature of the gas. Together, γ and t span a rich diagram of distinct phases (Fig. 2.1). The strongly interacting regime discussed above survives for low temperatures with $t \ll 1$. At high temperatures, interactions do not play a significant role compared to the thermal energy and the system can be described as a nearly ideal Bose gas, with none of the $T = 0$ phases occurring. The crossover to the quasi-condensate regime happens at $t\gamma^{3/2} = 1$ where interactions start to become important and suppress the fluctuations in density.

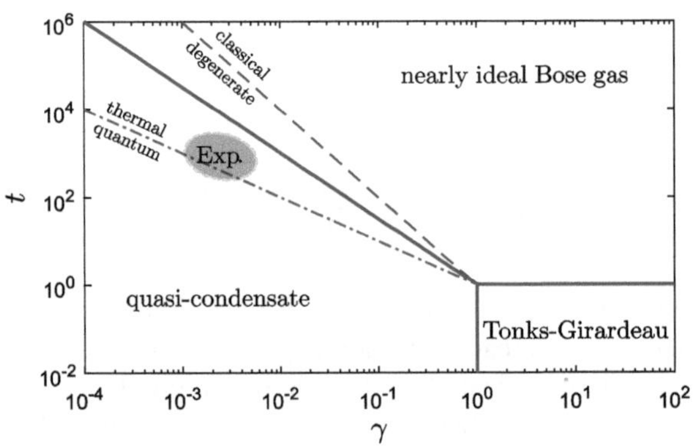

Fig. 2.1 Phases of the Lieb-Liniger model. Diagram showing the finite temperature phases of the 1d Bose gas with contact interactions. Separated by solid lines are the Tonks-Girardeau regime ($\gamma \gg 1,\ t \ll 1$), the nearly ideal Bose gas ($t\gamma^{3/2} > 1,\ t \gg 1$) and the quasi-condensate regime ($t\gamma^{3/2} < 1,\ \gamma \ll 1$). Furthermore, the dashed line indicates the degeneracy transition ($t\gamma^2 = 1$) and the dash-dotted line marks the thermal-quantum boundary ($t\gamma = 1$) for the fluctuations in the quasi-condensate regime. All boundaries between phases are smooth crossovers rather than sharp transitions. The gray shaded ellipse indicates the parameter regime of the measurements discussed in this thesis

At the border between the ideal Bose gas and the quasi-condensate regime some interesting subregimes can be identified. For the ideal Bose gas, around the point where the thermal de Broglie wavelength gets on the order of the inter-particle spacing $n_{1\mathrm{d}}\lambda_{\mathrm{dB}} \approx 1$, the gas becomes degenerate and can no longer be described by a Maxwell-Boltzmann distribution. From $\lambda_{\mathrm{dB}} = \sqrt{2\pi\hbar^2/mk_{\mathrm{B}}T}$ we see that this crossover appears at $t\gamma^2 = 1$. Further, in the quasi-condensate regime close to the ideal gas transition mode occupations are large and thermal fluctuations dominate the gas. Only for $t\gamma \ll 1$ quantum fluctuations become important.

The experiments discussed in this thesis cover a small portion of this varied phase space, as indicated in Fig. 2.1. They approximately span $1 \cdot 10^{-3} < \gamma < 5 \cdot 10^{-3}$ and $2 \cdot 10^2 < t < 2.5 \cdot 10^3$ and mainly fall in the quasi-condensate phase dominated by thermal fluctuation. This regime will therefore be discussed in more detail in the following section.

Note that the regimes discussed here are not separated by sharp transitions but by smooth crossovers. For example, although the experimentally covered parameter space extends slightly into the quantum quasi-condensate regime (see Fig. 2.1) the results discussed in this thesis can generally be explained by classical fields models. This is because the quantum-thermal border signifies the point where the modes around $k\xi \approx 1$, with $\xi = \hbar/\sqrt{2mg_{1\mathrm{d}}n_{1\mathrm{d}}}$ being the healing length,[3] are no longer macroscopically occupied. However, the low-energy modes dominating the experimental observables get quantum only at lower temperatures currently inaccessible in our setup [21].

2.1.2 Effective Description for Quasi-condensates

Although analytic solutions exist for the entire phase space of the interacting 1d Bose gas they are not practical to calculate dynamics. Therefore, in order to describe the experimental system and get an intuition for the underlying physics in the accessible parameter regime we turn to effective models. These models are devised to describe the low-energy collective dynamics of the gas, aligning well with the experimentally accessible observables. For single quasi-condensates such an effective description is found in reference [22] and we will discuss the assumptions and implications of this approach in the following.

The goal is to do a perturbative expansion of the governing Hamiltonian and collect the leading terms describing our system. For a weakly interacting condensate in 3d this can be done by writing the field operator as a combination of the macroscopically occupied mode ϕ_0 and small perturbations on top of that mode, $\hat{\psi} = \phi_0\,\hat{a}_0 + \delta\hat{\psi}$. Here, $\delta\hat{\psi}$ contains all excitations, be they quantum or thermal, and $\hat{a}_0$ is the annihilation operator for the condensate mode. Assuming the occupation of $\hat{a}_0$ to be much larger than that of the excitations, the Hamiltonian can be expanded in

[3]This is the length scale on which a condensate wave function returns to its bulk value next to a local perturbation [14].

the small parameter $\delta\hat{\psi}$. For quasi-condensates however, no single macroscopically occupied mode exists such that the phase of the condensate fluctuates strongly. This absence of long range phase order renders the approach described above incompatible. Yet, one can still rely on the suppression of density fluctuations, motivating the expression of the field operator in a phase-density representation

$$\hat{\psi}(z,t) = e^{i\hat{\theta}(z,t)}\sqrt{n_{1d}(z) + \delta\hat{n}(z,t)}\ , \tag{2.5}$$

where $\hat{\theta}(z,t)$ describes the fluctuating phase and $\delta\hat{n}(z,t)$ the density fluctuations relative to the average density profile $n_{1d}(z)$. The two new operators inherit the commutation relations from the bosonic field, such that $[\delta\hat{n}(z), \hat{\theta}(z')] = i\delta(z-z')$. Note however that this definition of the phase operator hides some subtleties and only makes sense in a coarse grained description where each bin shows a finite occupation of particles. Nevertheless, for clarity we will always assume the continuum limit in this discussion and refer to Ref. [22] for details.

The Hamiltonian of interest is given by Eq. (2.2) but defined in the grand canonical ensemble with the chemical potential μ setting the total particle number. Further, in order to describe the experimentally realized system we add the confining external potential $U(z)$ leading to

$$\hat{H} = \int \mathrm{d}z\, \hat{\psi}^\dagger(z)\left[-\frac{\hbar^2}{2m}\partial_z^2 + U(z) - \mu + \frac{g_{1d}}{2}\hat{\psi}^\dagger(z)\hat{\psi}(z)\right]\hat{\psi}(z)\ . \tag{2.6}$$

Being interested in the low-energy dynamics we can assume $|\delta\hat{n}|/n_{1d} \ll 1$ and $|\partial_z\hat{\theta}|/n_{1d} \ll 1$, corresponding to small density fluctuations and long wavelength phase fluctuations. Inserting the ansatz of Eq. (2.5) into Eq. (2.6) and expanding the Hamiltonian up to second order in these small parameters[4] reveals the leading contributions: $\hat{H} = \hat{H}^{(0)} + \hat{H}^{(1)} + \hat{H}^{(2)} + \dots$. The zeroth order term $\hat{H}^{(0)}$ only depends on the classical background density profile $n_{1d}(z)$ and is minimized by the solution of the Gross-Pitaevskii equation (GPE)

$$\left[-\frac{\hbar^2}{2m}\partial_z^2 + U(z) - \mu + g_{1d}n_{1d}(z)\right]\sqrt{n_{1d}(z)} = 0\ . \tag{2.7}$$

The first order correction $\hat{H}^{(1)}$ vanishing for $n_{1d}(z)$ fulfilling the GPE above, leaving the second order corrections to dominate the fluctuations in the system with[5]

[4]Assuming both expansion parameters to be of the same size.

[5]Whenever convenient we are suppressing the z and t dependence of $\delta\hat{n}$, $\hat{\theta}$, n_{1d} and the quantities derived from them to simplify the notation and aid readability.

$$\hat{H}^{(2)} = \int \mathrm{d}z \Big[-\frac{\hbar^2}{8m}\frac{\delta\hat{n}}{\sqrt{n_{1\mathrm{d}}}}\partial_z^2\left(\frac{\delta\hat{n}}{\sqrt{n_{1\mathrm{d}}}}\right) + \\ + \frac{\hbar^2}{2m}\left(\frac{\delta\hat{n}^2}{4n_{1\mathrm{d}}^2} + \hat{\theta}^2\right)\sqrt{n_{1\mathrm{d}}}\partial_z^2\sqrt{n_{1\mathrm{d}}} + \\ + \frac{g_{1\mathrm{d}}}{2}\delta\hat{n}^2 - \frac{\hbar^2}{2m}\sqrt{n_{1\mathrm{d}}}\hat{\theta}\,\partial_z^2\left(\sqrt{n_{1\mathrm{d}}}\hat{\theta}\right)\Big] . \tag{2.8}$$

Reference [22] shows how this Hamiltonian can be diagonalized by inserting Eq. (2.7) into its second term and performing a Bogoliubov transformation. Here, however, we will do some further approximations restricting ourselves to the the experimentally relevant lowest energy modes. For these long wavelength excitations accessible in the experiment the first term in Eq. (2.8), generally called the quantum pressure term, can be neglected over the third term. Also, in a Thomas-Fermi approximation, neglecting the small kinetic energy of the background density profile in Eq. (2.7), the second term in Eq. (2.8) can be neglected [23]. Further, in the fourth term $\sqrt{n_{1\mathrm{d}}}$ can be pulled out of the derivative. This leaves the quadratic Hamiltonian[6]

$$\hat{H}_{LL} = \int \mathrm{d}z \left[\frac{g_{1\mathrm{d}}}{2}\delta\hat{n}^2 + \frac{\hbar^2 n_{1\mathrm{d}}}{2m}\left(\partial_z\hat{\theta}\right)^2\right] , \tag{2.9}$$

which represents a realization of the Luttinger liquid model [24–26]. Initially devised to describe interacting electrons in 1d this model constitutes the low energy description of a large class of 1d problems [27, 28]. Its discussion will lay the groundwork for much of the physics presented in this thesis.

Even though the external potential does not explicitly enter Eq. (2.9), it shapes the density profile $n_{1\mathrm{d}}(z)$ through the GPE in Eq. (2.7) and thereby the eigenmodes of the system. In the following we will discuss different geometries starting with the simple case of a homogeneous system of length L enclosed by infinitly hard walls. As the walls of this ideal box force the particle flux at the boundary to zero the phase field, whose spatial derivative is proportional to the flux, has to fulfill Neumann boundary conditions

$$\partial_z\hat{\theta}(z)\big|_{z=0,\,L} = 0 . \tag{2.10}$$

This allows for modes with wave numbers $k = \frac{\pi}{L}j$ where $j = 1, 2, 3, \ldots$ is the mode index, as illustrated in Fig. 2.2.[7] Expanding the fluctuations in these modes we obtain

$$\delta\hat{n}(z) = \sqrt{\frac{2}{L}}\sum_k \delta\hat{n}_k \cos(kz) , \quad \hat{\theta}(z) = \sqrt{\frac{2}{L}}\sum_k \hat{\theta}_k \cos(kz) , \tag{2.11}$$

[6] For the forth term we used $-\int \mathrm{d}z\,\hat{\theta}\partial_z^2\hat{\theta} = \int \mathrm{d}z\,(\partial_z\hat{\theta})^2$.

[7] In the case of periodic boundary conditions only every second of these frequencies would be supported due to the continuity conditions $\hat{\theta}(0) = \hat{\theta}(L)$ and $\delta\hat{n}(0) = \delta\hat{n}(L)$.

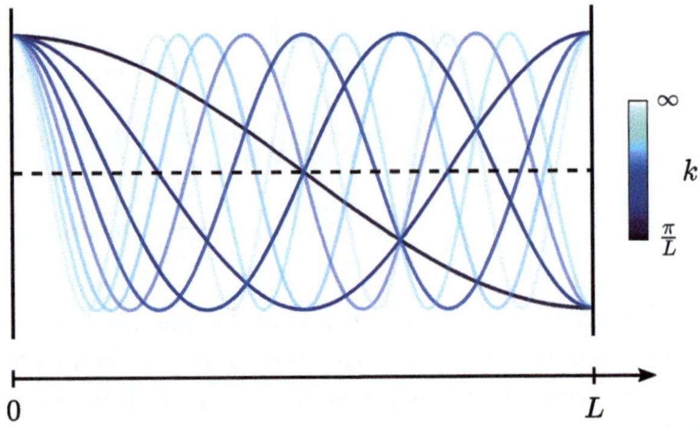

Fig. 2.2 Phase fluctuation modes in the ideal box. Illustration of the box modes supported by the Neuman boundary conditions imposed through the hard walls

with $\delta\hat{n}_k$ and $\hat{\theta}_k$ being the mode amplitudes fulfilling the commutation relation $[\delta\hat{n}_k, \hat{\theta}_{k'}] = i\delta_{k,k'}$. Inserting these expansions into Eq. (2.9) with the integration performed over $z \in [0, L]$ results in the momentum space representation of the Luttinger liquid Hamiltonian

$$\hat{H}_{LL} = \sum_{k\neq 0}\left[\frac{g_{1\text{d}}}{2}\,\delta\hat{n}_k^2 + \frac{\hbar^2 k^2 n_{1\text{d}}}{2m}\,\hat{\theta}_k^2\right] + \hat{H}_0\,. \tag{2.12}$$

The term $\hat{H}_0$ corresponds to the $k = 0$ mode and will be neglected in the following as it does not enter most observables. From Eq. (2.12) we can already see that these modes do not couple as cross terms between different k modes do not exist. Defining creation and annihilation operators $\hat{b}_k^\dagger$, $\hat{b}_k$ for each mode we can construct the mode amplitudes as

$$\delta\hat{n}_k = \sqrt{\frac{n_{1\text{d}}}{2}\frac{\epsilon_k}{\mu}}\left(\hat{b}_k\,e^{-i\omega_k t} + \hat{b}_k^\dagger\,e^{i\omega_k t}\right),$$
$$\hat{\theta}_k = -i\sqrt{\frac{1}{2n_{1\text{d}}}\frac{\mu}{\epsilon_k}}\left(\hat{b}_k\,e^{-i\omega_k t} - \hat{b}_k^\dagger\,e^{i\omega_k t}\right),$$

leading to the diagonal form of the Hamiltonian

$$\hat{H}_{LL} = \sum_k \hbar\omega_k\,\hat{b}_k^\dagger\hat{b}_k\,. \tag{2.13}$$

The mode energy $\epsilon_k = \hbar\omega_k$ is given by the linear dispersion relation

$$\omega_k = ck = \frac{c\pi}{L}j\,, \tag{2.14}$$

with $c = \sqrt{g_{1\text{d}}n_{1\text{d}}/m}$ being the speed of sound. This shows the low-energy dynamics of the gas to be dominated by free phononic excitations whose energy is equally spaced. The zero point energy resulting from the commutator $[\hat{b}_k, \hat{b}_k^\dagger] = \delta_{k,k'}$ is already subtracted in Eq. (2.13).

As discussed in Sect. 2.1.1 the experimentally accessible parameter regime is restricted to temperatures where the average occupation of the observable low lying modes is much larger than one, $\langle \hat{b}_k^\dagger \hat{b}_k \rangle \gg 1$. For the rest of this manuscript we will therefore treat $\delta n(z)$ and $\theta(z)$ as classical fields. In this approximation a thermal state displays the occupations $\langle b_k^\dagger b_k \rangle = k_\mathrm{B} T/\epsilon_k$. This equipartition of energy among modes further extends to the phase and density fluctuation amplitudes within each mode such that

$$\langle |\delta n_k|^2 \rangle = \frac{k_\mathrm{B} T}{g_\mathrm{1d}} , \qquad \langle |\theta_k|^2 \rangle = \frac{m k_\mathrm{B} T}{\hbar^2 k^2 n_\mathrm{1d}} . \tag{2.15}$$

The spatial correlations for such a thermal state will be discussed in Sect. 2.1.3 as the configuration of two adjacent gases allows for their direct observation (see Sect. 3.3.4).

Although seemingly removed from application the description of the gas in an ideal box describe many of the observations made in the imperfect experimental box trap discussed in Sect. 3.2.3 (see Chap. 5). The second relevant confining geometry is that of the harmonic trap, with $U(z) = \frac{1}{2} m \omega_z^2 z^2$. There, the Thomas-Fermi approximation of Eq. (2.7) leads to the inverted parabola profile

$$n_\mathrm{1d}(z) = n_0 \left[1 - \left(\frac{z}{R} \right)^2 \right] \Theta\left(R - |z| \right) ,$$

$$n_0 = \frac{3}{4} \left(\frac{2m\,\omega_z^2 N^2}{g_\mathrm{1d}} \right)^{1/3} , \quad R = \left(\frac{3N g_\mathrm{1d}}{2m\omega_z^2} \right)^{1/3} \tag{2.16}$$

with n_0 being the density in the center of the trap, R the radius of the cloud, N the number of particles in the trap and $\Theta(z)$ the Heaviside function. As in Eq. (2.11) we can expand the fluctuations in the eigenmodes of the Hamiltonian

$$\delta n(z) = \sum_j \delta n_j\, f_j(z) , \quad \theta(z) = \sum_j \theta_j f_j(z) , \tag{2.17}$$

with the mode functions $f_j(z)$ now being defined by Legendre polynomials, $f_j(z) = \sqrt{j + 1/2} P_j(z/R)$ [29]. Proceeding as for the ideal box, defining creation and annihilation operators for each mode, we arrive at a diagonal Hamiltonian with the mode energies

$$\omega_j = \omega_z \sqrt{j(j+1)/2} . \tag{2.18}$$

Compared to the homogeneous case this dispersion is neither equally spaced nor commensurate which will become important in Chap. 5 where we discuss the dynamic recurrence of the system due to a rephasing of excitations.

With the quantum pressure term in Eq. (2.8) neglected the above descriptions are valid for $k\xi \ll 1$. However, when taking the term into account we can still diagonalize the Hamiltonian through free excitations and push the validity beyond these energies [22]. For the homogeneous case the mode energies are then given by

$$\omega_k = \sqrt{\frac{\hbar^2 k^2}{2m}\left(\frac{\hbar^2 k^2}{2m} + 2\mu\right)} .$$

For $k\xi \ll 1$ this reduces to the linear dispersion found above while for $k\xi \gg 1$ we recover the free particle dispersion $\omega_k = \hbar^2 k^2/2m$. Due to most of the experimental observables being dominated by low-k modes we can safely restrict ourselves to the linear phononic dispersion for most discussions in this thesis.

Both the Luttinger liquid and the Bogolibov description taking the quantum pressure term into account are integrable, their trivial conserved quantities being the eigenmode occupations. Going beyond second order in the expansion of Eq. (2.6) breaks this integrability and leads to terms mediating interactions between the modes. This does not contradict the fact that the Lieb-Liniger model, at least in the homogeneous case, is integrable. The eigenmodes of the effective description simply do not coincide with the conserved charges of the Lieb-Liniger Hamiltonian. Therefore, integrable dynamics on the level of the full description can present itself as an apparent thermalization in the basis of the effective descriptions.

2.1.3 *Coupled Quasi-condensates*

A key feature of our setup is the ability to realize a double well potential along one of the tightly confined dimensions of the trap (see Sect. 3.1.3). This allows the realization of two adjacent 1d Bose gases separated by an adjustable barrier. As in the case of the single gas, if the internal energy scales are low enough we can restrict the description to the lowest transverse states in each of the wells.[8] This means each gas can be described by the Hamiltonian of a single gas (Eq. 2.6) while additionally experiencing a coupling from particles tunneling through the inter-well barrier. If the barrier is large enough and the overlap between the two wave functions is small, such linear coupling dominates and interactions between the wells can be neglected. The Hamiltonian of the entire system then reads

$$\begin{aligned} H &= H_1 + H_2 + H_J \\ &= \sum_{i=1,2} \int \mathrm{d}z\, \psi_i^\dagger(z) \left[-\frac{\hbar^2}{2m}\partial_z^2 + U(z) - \mu + \frac{g_{1\mathrm{d}}}{2}\psi_i^\dagger(z)\psi_i(z) \right] \psi_i(z) - \\ &\quad - \int \mathrm{d}z\, \hbar J \left(\psi_1^\dagger(z)\psi_2(z) + \psi_2^\dagger(z)\psi_1(z) \right) , \end{aligned}$$

[8] This is equivalent to the two mode approximation of the double well potential [30].

with $i = 1, 2$ being the well index and J being the coupling strength which can be tuned by changing the overlap between the transverse wave functions, i.e. raising or lowering the barrier between the wells.

We can now apply the same treatment as for the single gas in Sect. 2.1.2, writing the wave functions in phase-density representation and performing a perturbative expansion in $\delta n_{1,2}(z)$ and $\partial_z \theta_{1,2}(z)$. Applying the same approximations as before, for $H_{1,2}$ this leads to two independent Luttinger liquid Hamiltonians of the form given in Eq. (2.9). Before considering the coupling term H_J, however, it is convenient do a variable transformation to common (c) and relative (r) fluctuations

$$\begin{aligned} \delta n_c(z) &= \delta n_1(z) + \delta n_2(z)\,, & \varphi_c(z) &= \frac{1}{2}[\theta_1(z) + \theta_2(z)]\,, \\ \delta n_r(z) &= \frac{1}{2}[\delta n_1(z) - \delta n_2(z)]\,, & \varphi_r(z) &= \theta_1(z) - \theta_2(z)\,. \end{aligned} \tag{2.19}$$

This is helpful as the coupling term predominantly affects the relative degrees of freedom and since the relative phase field $\varphi_r(z)$ constitutes our main experimental observable. The normalization of the transformation is in principle arbitrary but chosen such that $\varphi_r(z)$ best aligns with the measured quantity (see Sect. 3.3.4). Transforming the Luttinger liquid Hamiltonians in each well to the new degrees of freedom results again in two independent Luttinger liquids for the common and relative fluctuations, respectively, $H_{LL,1} + H_{LL,2} = H_{LL,c} + H_{LL,r}$. Note however that this is only the case if the average density in both wells is the same. For an imbalanced double well an additional term proportional to the density difference $n_{1\mathrm{d},1} - n_{1\mathrm{d},2}$ arises, coupling common and relative degrees of freedom [31]. In the following discussion we will assume the wells to be perfectly balanced, although experimentally this is never the case. Later in Chap. 5 we will discuss the common-relative coupling induced by an imbalance spread due to a distribution of experimental initial conditions.

Turning to the actual tunneling coupling and expanding H_J up to second order we obtain

$$H_J^{(2)} = -2\hbar J\, n_{1\mathrm{d}} \cos(\varphi_r) \left[1 + \frac{\delta n_c}{2n_{1\mathrm{d}}} - \frac{\delta n_r^2}{2n_{1\mathrm{d}}^2} \right]. \tag{2.20}$$

Here, the second term in the angled brackets is coupling common and relative degrees of freedom. However, in thermal equilibrium and at low energies this term is expected to have little influence [32]. Though it might be important in a non-equilibrium context we neglect it here as our focus lies in the equilibrium properties of the coupled system. Further approximating the last term in the angled brackets of Eq. (2.20) to contribute only to the δn_r^2 term in $H_{LL,r}$, we obtain for the Hamiltonian of the relative degrees of freedom

$$H_r = \int \mathrm{d}z \left[\tilde{g}_{1\mathrm{d}} \delta n_r^2 + \frac{\hbar^2 n_{1\mathrm{d}}}{4m} (\partial_z \varphi_r)^2 - 2\hbar J n_{1\mathrm{d}} \cos(\varphi_r) \right], \tag{2.21}$$

with $\tilde{g}_{1d} = g_{1d} + \hbar J/n_{1d}$. In this low-energy description common and relative fluctuations completely decouple and, as we are mainly interested in the relative dynamics, we will drop the r-subscripts from the phase and density fields from now on.

The Hamiltonian obtained in Eq. (2.21) represents a realization of the sine-Gordon model, a seminal field theory important in many branches of physics [33]. It is of particular interest as the cosine term brings it away from the somewhat trivial form of quadratic Hamiltonians, leading to non-Gaussian fluctuations and soliton excitations of the relative phase field. For equilibrium states it could be shown experimentally that it indeed describes the fluctuations between two coupled quasi-condensates [32]. However, for the experiments discussed in this thesis we are mainly interested in two limiting cases: vanishing coupling and strong coupling. In the first case, for $J = 0$, Eq. (2.21) reduces to the Luttinger liquid Hamiltonian of a single gas Eq. (2.9), up to prefactors stemming from the normalization in Eq. (2.19). The same diagonalization procedure can therefore be performed leading to the same linear dispersion relation. For strong coupling, the cosine term dominates the phase and localizes it around zero such that $\langle \cos(\varphi) \rangle \approx 1$. With the phase being a small quantity the cosine can be expanded keeping only the leading quadratic term which results in[9]

$$H_r = \int \mathrm{d}z \left[\tilde{g}_{1d} \delta n^2 + \frac{\hbar^2 n_{1d}}{4m} (\partial_z \varphi)^2 + \hbar J n_{1d} \varphi^2 \right] . \tag{2.22}$$

Again a quadratic Hamiltonian, H_r can be diagonalized in a straightforward way. For the ideal box confinement the relative phase field inherits the boundary conditions Eq. (2.10) from the individual gases and the fluctuations can be expanded in the same cosine modes as in (Eq. 2.11). However, the expansion coefficients change to

$$\begin{aligned} \delta n_k &= \sqrt{\frac{n_{1d}}{2}} \left(\frac{E_k + 2\hbar J}{2\mu + 2\hbar J} \right)^{1/4} \left(b_k \, e^{-i\omega_k t} + b_k^\dagger \, e^{i\omega_k t} \right) , \\ \varphi_k &= -i \sqrt{\frac{1}{2n_{1d}}} \left(\frac{2\mu + 2\hbar J}{E_k + 2\hbar J} \right)^{1/4} \left(b_k \, e^{-i\omega_k t} - b_k^\dagger \, e^{i\omega_k t} \right) , \end{aligned} \tag{2.23}$$

with $E_k = \frac{\hbar^2 k^2}{2m}$ and a dispersion relation given by

$$\epsilon_{k,J} = \sqrt{(E_k + 2\hbar J)(2\mu + 2\hbar J)} . \tag{2.24}$$

Here, we can see that the coupling opens a gap in the otherwise linear spectrum with $\epsilon_{k=0,J} \approx 2\sqrt{\hbar J \mu}$ for $\hbar J \ll \mu$ (Fig. 2.3a). At the same time the common degrees of freedom still retain the linear dispersion of a single gas.

---

[9] The constant first term being absorbed by a shift of the energy minimum that does not affect the dynamics.

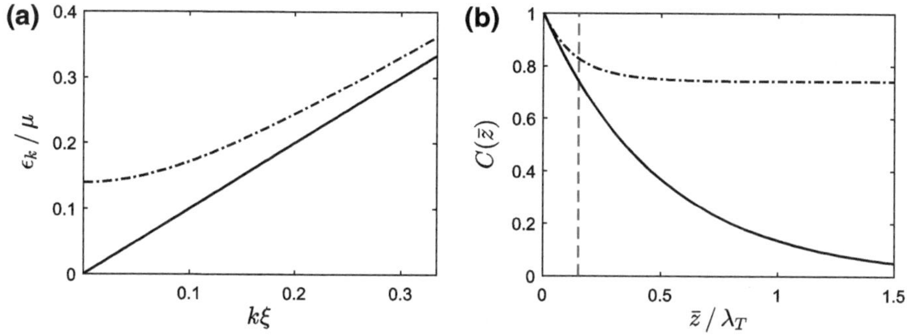

Fig. 2.3 Relative fluctuations in coupled quasi-condensates. **a** Linear dispersion of the relative modes between two uncoupled quasi-condensates (solid line) compared to the dispersion in the coupled case given by Eq. (2.24) (dash-dotted line). The parameters are set to values typical for the measurements presented in Chap. 5 ($n_{1d} = 70\ \mu m^{-1}$, $c = 2.2\ \mu m/ms$, $J = 2\pi \cdot 5\,Hz$). **b** Thermal phase scorrelation functions of the relative phases between two uncoupled gases given by Eq. (2.27) (solid line) and two strongly coupled gases with $l_J = 0.15 \cdot \lambda_T$ given by Eq. (2.28) (dash-dotted line). The dashed gray line indicates l_J. Both plots give the results obtained from the harmonic approximation made in Eq. (2.22)

Just as discussed for the single gas the effective model described by Eq. (2.22) is integrable and going beyond second order in the expansion around small fluctuations leads to terms that break this integrability. This forces the excitations discussed above to acquire a finite lifetime [34–36] and leads to a coupling of common and relative degrees of freedom. As these additional terms are of higher order they will only have significant effects over longer times, leaving the short time dynamics well described by Eq. (2.22).

Experimentally, our window into the the relative degrees of freedom is the phase field $\varphi(z)$ (see Sect. 3.3.4). Being able to measure the entire phase field of single realizations brings with it the possibility to evaluate spatial correlation functions. An intuitive form of such a correlation function is given by

$$C(z, z') = \left\langle \cos\left[\varphi(z) - \varphi(z')\right]\right\rangle . \tag{2.25}$$

This quantity, referred to as phase correlation function from now on, measures the correlations of φ at points z and z', giving 1 for perfectly correlated phases and zero for uncorrelated ones. It obtains further meaning when evaluating the two-point correlation function of the fields in both wells. The real part of this correlation function gives Eq. (2.25) when assuming density fluctuations to be negligible

$$\frac{\langle \psi_1(z)\, \psi_2^\dagger(z)\, \psi_1^\dagger(z')\, \psi_2(z')\rangle}{\langle |\psi_1(z)|^2\rangle \langle |\psi_2(z')|^2\rangle} \simeq \left\langle e^{i\left[\varphi(z)-\varphi(z')\right]}\right\rangle .$$

With the Hamiltonians for the uncoupled and the strongly coupled case both being quadratic we can assume the fluctuations always to be Gaussian. For a homogeneous system this leads to

$$\begin{aligned} C(\bar{z} = |z - z'|) &= \exp\left[-\frac{1}{2}\langle[\varphi(z) - \varphi(z')]^2\rangle\right] \\ &= \exp\left[-\int_0^\infty \frac{\mathrm{d}k}{\pi} \langle|\varphi_k|^2\rangle \; [1 - \cos(k\bar{z})]\right] . \end{aligned} \tag{2.26}$$

Inserting the thermal expectation value of φ_k for the uncoupled case and performing the integration this simplifies to

$$C(\bar{z}) = \exp\left[-\frac{2\bar{z}}{\lambda_T}\right] , \tag{2.27}$$

with $\lambda_T = 2\hbar^2 n_{1\mathrm{d}}/mk_{\mathrm{B}}T$ being the thermal coherence length of the gases (Fig. 2.3b). The factor of 2 in the exponent stems from the definition of the relative phase in Eq. (2.19) and does not appear for the phase correlations in a single gas. In the strongly coupled case we obtain

$$C(\bar{z}) = \exp\left[-\frac{2l_J}{\lambda_T}\left(1 - e^{-\frac{\bar{z}}{l_J}}\right)\right] , \tag{2.28}$$

for the correlation function of a thermal state. Here, $l_J = \sqrt{\hbar/4mJ}$ is the phase locking length scale. Beyond this length, long range phase fluctuations are suppressed by the coupling, leading to a flat correlation function (Fig. 2.3b). For the strong coupling limit where $l_J \ll \lambda_T$, this forces the phase field to be well localized around zero.

In Chap. 5 we will discuss the dynamics resulting from quenching the system from a strongly coupled initial state to an uncoupled situation by a fast ramp up of the double well barrier.

2.2 From 3d to 1d

As our world is made up of three spatial dimensions, any experimental realization of lower dimensional systems in the lab needs to either restrict or decouple the dynamics in the supernumerous dimensions. In cold atoms this is usually achieved by a tight confinement along these dimensions [37]. By bringing the internal energy scales of the gas far below the energy of the lowest excited states along these tightly confined dimensions all dynamics is frozen out and the system becomes effectively lower dimensional. For a 1d system in a tight harmonic potential with trap frequency $\omega_\perp$ this results in the 1d condition

$$\hbar\omega_\perp \gg k_B T,\ \mu\ . \tag{2.29}$$

If it is fulfilled, both the thermal energy and the interaction energy per particle are too low to allow for transverse excitations, forcing all particles to occupy the transverse ground state and the system becoming effectively 1d. In this section we will discuss the influence the transverse confinement has on the 1d dynamics and investigate the consequences of Eq. (2.29) not being strictly fulfilled.

2.2.1 *Effective 1d Interaction Strength*

Let us assume a gas confined in a tight harmonic potential along x and y and an arbitrary potential $U(z)$ along z, described by the mean field Hamiltonian

$$H = \int \mathrm{d}\mathbf{r}\ \Psi^*(\mathbf{r}) \left[-\frac{\hbar^2}{2m}\partial_\mathbf{r}^2 + \frac{m\omega_\perp^2(x^2+y^2)}{2} + U(z) + \frac{g_{3d}}{2}|\Psi(\mathbf{r})|^2 \right] \Psi(\mathbf{r})\ . \tag{2.30}$$

Interested only in the dynamics along z, the tightly confined dimensions can be integrated out, assuming the separability of the wave function $\Psi(\mathbf{r}) = \phi(x, y) \cdot \psi(z)$. Further neglecting interaction along x and y, for a cold gas we can expect all atoms to occupy the single particle ground state of the tight confinement, given by a Gaussian localized at the harmonic oscillator length $a_\perp = \sqrt{\hbar/m\omega_\perp}$,

$$\phi(x, y) = \frac{1}{a_\perp^2\sqrt{\pi}} e^{-\frac{(x^2+y^2)}{2a_\perp^2}}\ . \tag{2.31}$$

Integrating Eq. (2.30) along x and y leads to a constant energy shift[10] $\hbar\omega_\perp$ irrelevant for the dynamics and to a renormalization of the interaction strength

$$g_{1d} = g_{3d} \int \mathrm{d}x\, \mathrm{d}y\ |\phi(x, y)|^4 = 2\hbar\omega_\perp a_s. \tag{2.32}$$

with g_{3d} given by Eq. (2.1). Note that in contrast to the 3d case g_{1d} explicitly depends on the trap frequency of the transverse potential, enabling tuning of the interaction strength to a certain extent.

When $a_\perp$ becomes of the same order as a_s a more careful analysis of the two-body scattering process has to be performed. In this case, virtual transverse excitations lead to a confinement induced resonance [38, 39]

[10] This corresponds to the ground state energy of the 2d harmonic potential.

$$g_{1d} = 2\hbar\omega_\perp a_s \left(1 - \vartheta \frac{a_s}{a_\perp}\right)^{-1}, \tag{2.33}$$

where $\vartheta \approx 1$ is a numerical factor. For the transverse trap frequencies of $\omega_\perp = 2\pi \cdot 1.4 - 2.1\,\text{kHz}$ realized in our setup $a_\perp$ is much larger than a_s such that the scattering retains its 3d character and the correction in (2.33), amounting to 2%, can be neglected.

2.2.2 Transverse Broadening

Assuming all atoms to occupy the transverse single particle ground state given in Eq. (2.31), neglecting the effects of interactions along these dimensions, is only valid as long as the density is small $a_s n_{1d} \ll 1$. In the cross over regime for $a_s n_{1d} < 1$, a variational approach can be used to calculate corrections to the effective 1d interactions obtained in Eq. (2.32) [40]. As interactions are expected to lead to a broader state, a natural ansatz for the transverse wave function is a Gaussian of variable width, $\phi(x, y; \sigma(z,t))$. Again assuming separability of the 3d wave function, we can insert this ansatz into the Hamiltonian given in Eq. (2.30), integrate over x and y and minimize the action functional. Assuming that ϕ only varies slowly along z this gives the density dependent transverse width $\sigma^2 = a_\perp^2\sqrt{1 + 2a_s n_{1d}}$ and the chemical potential

$$\mu = \hbar\omega_\perp \frac{1 + 3a_s n_{1d}}{\sqrt{1 + 2a_s n_{1d}}} = \hbar\omega_\perp + g_{1d} n_{1d} + \frac{3}{4} g_{1d} a_s n_{1d}^2 + O\left[(a_s n_{1d})^3\right]. \tag{2.34}$$

As shown in the expansion on the right hand side, this expression includes the transverse ground state energy and gives the 1d chemical potential at first order in $a_s n_{1d}$. Comparing the results from this approach with GPE simulations of the full mean-field solution in the xy-plane shows good agreement in the transverse density distribution, the width and the chemical potential (Fig. 2.4a, b and c, respectively). Figure 2.4d shows how the interaction energy gets gradually more important when the density is increased. Going beyond the experimentally relevant densities this plot would show the cross-over to the Thomas-Fermi regime for the transverse wave function where the interaction energy approaches the potential energy and the kinetic term can be entirely neglected.

The equations of motion in the remaining longitudinal dimension obtained from this approach give a non-polynomial extension of the GPE

$$i\hbar\partial_t\psi(z,t) = \left[-\frac{\hbar^2}{2m}\partial_z^2 + U(z) + \hbar\omega_\perp \frac{1 + 3a_s|\psi(z,t)|^2}{\sqrt{1 + 2a_s|\psi(z,t)|^2}}\right]\psi(z,t)\,. \tag{2.35}$$

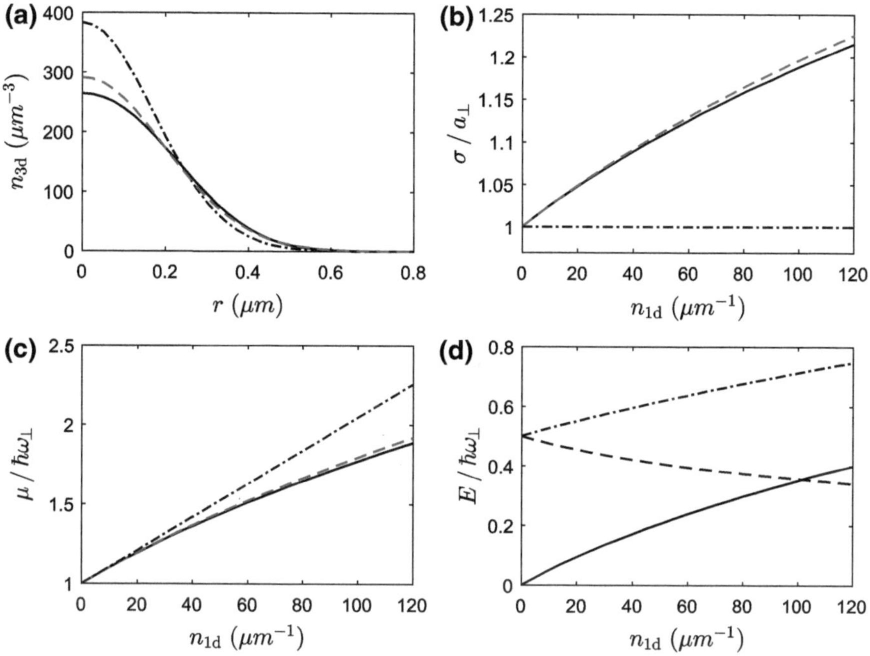

Fig. 2.4 Interaction effects on the transverse wave function. a Comparison of the transverse density profile obtained from imaginary time evolution of the radial GPE (solid line) (see Appendix B) with the single particle ground state (dash-dotted line) and the broadened Gaussian obtained from the variational approach (dashed gray line). All profiles are calculated for a linear density of $n_{1d} = 80\ \mu m^{-1}$. **b, c** Density dependence of the transverse width σ and the chemical potential μ for the states plotted in (a) using the same line styles. **d** Density dependence of the kinetic (dashed line), potential (dash-dotted line) and interaction energy (solid line) obtained from the radial GPE simulations

For an inhomogeneous longitudinal confinement the most obvious effect of the broadening is the distortion of the ground state density profile. In the typical harmonic potential $U(z) = m\omega_z^2 z^2/2$ the density distribution is squeezed, increasing the central density and reducing the system size compared to Eq. (2.16) (Fig. 2.5a). For more homogeneous potentials like the nearly ideal box discussed in Sect. 3.2.3 the effect is less pronounced (Fig. 2.5b).

Beyond the profile distortion the broadening has a marked effect on the excitation spectrum. Calculating the speed of sound from the the chemical potential given in Eq. (2.34) through the hydrodynamic relation [2, 41]

$$c = \sqrt{\frac{n_{1d}}{m} \frac{\partial \mu}{\partial n_{1d}}} = c_0 \sqrt{\frac{1}{2} \frac{2 + 3a_s n_{1d}}{(1 + 2a_s n_{1d})^{3/2}}} = \sqrt{\frac{g_n\, n_{1d}}{m}}, \tag{2.36}$$

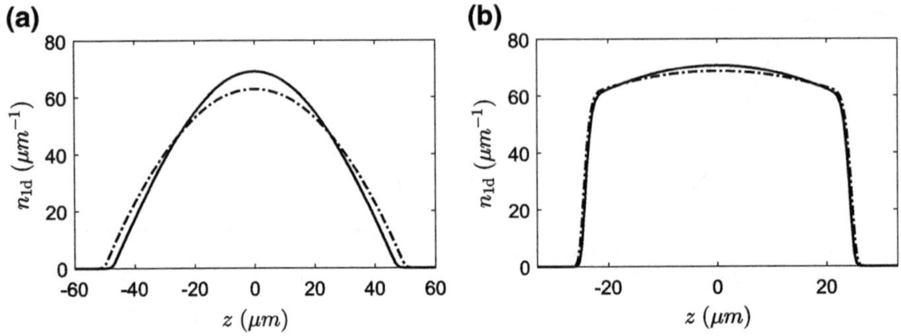

Fig. 2.5 Longitudinal ground state density under transverse broadening. Comparison of the ground state density profile obtained from the 1d GPE as given in Eq. (2.7) (dash-dotted line) and the the non-polynomial extension of the GPE given in Eq. (2.35) (solid line). **a** shows the case of a harmonic potential and **b** the nearly ideal box trap discussed in Sect. 3.2.3. All profiles are obtained through imaginary time evolution (see Sect. 2.3.2)

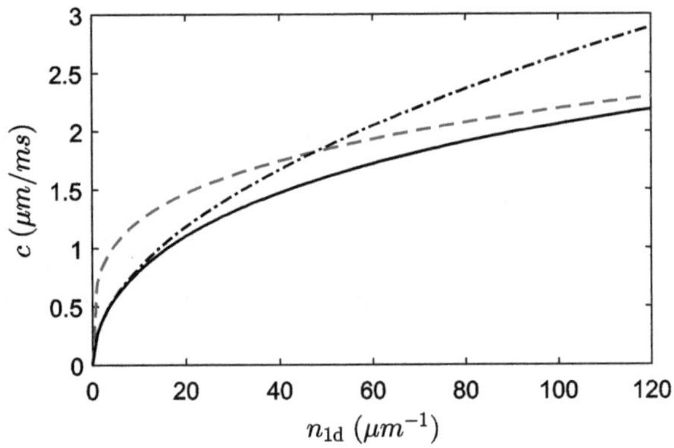

Fig. 2.6 Broadening effects on the speed of sound. Comparison of the bare speed of sound c_0 (dash-dotted line) with the broadened speed of sound given in Eq. (2.36) (solid line). The dashed gray line shows the 3d prediction which the broadened speed of sound approaches for large densities

we obtain a complicated density dependency. Here, $c_0 = \sqrt{g_{1d}n_{1d}/m}$ is the bare speed of sound introduced in Sect. 2.1.2 and g_n is the effective density dependent interaction strength emerging for the second order description of the fluctuations. The broadening effects on the speed of sound are by no means a small perturbation as shown in Fig. 2.6. For the typical densities of the measurements presented in Chap. 5 ($n_{1d} = 70\ \mu m^{-1}$) the change in c amounts to 20%, crucially influencing the timing of the return of coherence observed. Inserting g_n in the Luttinger liquid Hamiltonian given in Eq. (2.9) or the coupled model in harmonic approximation given in Eq. (2.22) allows us to take these broadening effects into account, especially in the numerical calculations discussed in Sect. 2.3.1.

It is interesting to note that the observed reduction in the speed of sound for higher densities reflects the transition to the 3d regime. Figure 2.6 shows that for typical experimental densities we are already closer to the 3d result than to the purely 1d prediction [41, 42].[11]

2.2.3 Collisions

With the 1d condition given in Eq. (2.29) strictly fulfilled elastic collisions between atoms are fully restricted to the single remaining dimension. In such collisions, energy and momentum conservation only allow for a swap of momenta between the particles and thereby forbid the redistribution of energy. This leads to the inhibition of thermaization in 1d Bose gases discussed earlier. However, in experimental implementations Eq. (2.29) will only be fulfilled up to a certain degree. If transverse excited states are populated with finite probabilities, e.g. due to $k_B T \not\ll \hbar\omega_\perp$, the 1d paradigm is broken and collisions involving these states can redistribute energy and lead to thermalization. The rate of such thermalizing two-body collisions was calculated in Ref. [43] to be

$$\Gamma_{2b} = 2\sqrt{2}\omega_\perp \frac{a_s^2 n_{1d}}{a_\perp} e^{-\frac{2\hbar\omega_\perp}{k_B T}} . \tag{2.37}$$

The exponential factor represents the Boltzmann weight of states with a high enough energy to facilitate transverse excitations, strongly suppressing the rate even at moderately low temperatures. For typical experimental parameters of the measurements presented in this thesis Γ_{2b} is on the order of a few Hertz or below, rendering the process much slower than the observed dynamics. Also, note that Eq. (2.37) was calculated for non-degenerate atoms and is expected to be even more suppressed for a degenerate gas [44].

However, even with the thermal population of excited transverse states strongly suppressed, their mere presence can open up new atomic scattering channels. Specifically, virtual transverse excitations can mediate effective three-body collisions [43, 45]. In such processes two atoms scatter, one gets virtually excited but gets immediately deexcited in a second two-body process involving a third atom, in total allowing for a redistribution of energy. In leading order the rate of these collisions is predicted to be independent of temperature, dominating over Γ_{2b} for cold samples. Yet, for our typical experimental parameter range we obtain $\Gamma_{3b} \approx 2\pi \cdot 0.5\,\text{Hz}$ which is also much slower than the dynamics we are interested in.

Interestingly, also the broadening effects discussed in Sect. 2.2.2 can be understood as an admixture of transverse excited single particle orbitals to the many-body ground state. Up to second order they only lead to a renormalization of the effective 1d interaction strength but going to third order one can show that a new term arises

[11] Note that Eq. (2.36) is not exactly converging to the 3d result $c_{3d} = \sqrt{\hbar\omega_\perp/m}(a_s n_{1d})^{1/4}$ due to the approximation made by taking the broadened Gaussian ansatz.

(see Appendix A). Adding this term to the Luttinger liquid Hamiltonian in Eq. (2.9) we obtain

$$H_{LL} + H^{(3)}_{\text{broad}} \simeq \int \mathrm{d}z \left[\frac{g_n}{2}\delta n^2 \left(1 - \frac{1}{2}a_s\delta n\right) + \frac{\hbar^2 n_{1\text{d}}}{2m}\left(\partial_z\hat{\theta}\right)^2\right]. \quad (2.38)$$

As shown shown in Ref. [43] the interaction term stemming from the virtual three-body collisions is (almost) equivalent to the first order cubic corrections of the broadening. Therefore, $H^{(3)}_{\text{broad}}$ describes the leading phonon-phonon interactions induced by the virtual atomic three-body collisions that break the purely 1d configuration. However, as already discussed for $\Gamma_{3\text{b}}$, with $a_s\delta n < a_s n_{1\text{d}} < 1$ being quite small this correction should not have significant effects on the intermediate time dynamics of our system.

2.3 Numerical Methods

Finally, in this section we will discuss some numerical tools that allow us to efficiently calculate the dynamics of inhomogeneous systems. We start with a numerical implementation of the Luttinger liquid model, including the coupling in harmonic approximation. It allows us to calculate thermal states and dynamical properties for arbitrary confinements beyond the ideal box and the harmonic potential discussed in Sect. 2.1.2. Furthermore, we shortly review a numerical implementation of the GPE and discuss how we obtain thermal initial states for our system.

2.3.1 Discretized Luttinger Liquid

To implement the mean field Hamiltonian describing the relative degrees of freedom between two coupled Luttinger liquids in harmonic approximation (Eq. 2.22) on an arbitrary background destiny we first discretize the problem casting the fluctuating fields $\vec{\varphi}$ and $\vec{\delta n}$ into vectors on the spatial grid. As phase and density fluctuations completely decouple we can write the density matrix of a thermal state with inverse temperature $\beta = 1/k_\text{B}T$ as

$$\varrho = \frac{1}{Z}\exp[-\beta H] = \frac{1}{Z}\exp\left[-\frac{\beta}{2}\left(\vec{\delta n}^T L_{LL}\vec{\delta n} + \vec{\varphi}^T\left(K_{LL} + K_J\right)\vec{\varphi}\right)\right],$$

where $Z = \text{Tr}\left[\exp(-\beta H)\right]$ is the partition function, L_{LL} and K_{LL} are matrices that represent the discretized Luttinger liquid Hamiltonian (first two terms of Eq. (2.22), respectively) and the matrix K_J implements the tunneling coupling in quadratic approximation (last terms of Eq. (2.22)). These matrices include the spatially varying background density $n_{1\text{d}}(z)$ as well as the density dependent interaction constant g_n

resulting from the broadening of the transverse wave function (see Sect. 2.2.2). In mean field, since ϱ is a product of two multivariate Gaussian distributions, the variances of the phase and density fluctuations can simply be obtained from inverting $\beta(K_{LL} + K_J)$ and βL_{LL}. Thereby, thermal states for any density profile or coupling can be obtained. Setting K_J to zero and adapting the prefactors to describe Eq. (2.9), also the fluctuations within a single gases can be described. Dynamics can further be calculated from the fields Heisenberg equations of motion, for any initial state. The employed code used to obtain initial states and propagate them was written by Thomas Schweigler [46].

2.3.2 *Gross-Pitaevskii Equation*

Going beyond the quadratic description we can numerically solve the non-polynomial GPE given in Eq. (2.35). This provides a mean field description that also includes terms mediating scattering between the phonon modes that diagonalize the Luttinger liquid model. For the implementation we use a standard second order split-step Fourier spectral method [47] which allows for the time propagation of arbitrary initial states and the calculation of ground states through an imaginary time evolution.

However, in order to faithfully describe the experimental dynamics it is crucial to start from thermal initial states. Whereas for 3d GPE simulations it is often sufficient to start from a state at a defined energy that thermalizes during some initial time evolution, the suppressed thermalization in 1d precludes such procedures [48]. A common approach to obtain (nearly) thermal initial states in homogeneous 1d systems is to populate the Bogoliubov modes according to a thermal distribution. Yet, this demands knowledge of the mode functions which are not directly available for arbitrary trapping geometries. For the simulations discussed here we therefore opt for an initialization through a stochastic GPE (SGPE) [49]. There, a thermal state is obtained from modeling the contact of the system to a thermal background reservoir through incoherent scattering by adding noise and dissipation terms to the regular GPE. The resulting Langevin equation for a single gas is given by

$$i\hbar\, \partial_t \psi = (1 + i\gamma)\, H_{\mathrm{GP}}\, \psi + \eta\,, \tag{2.39}$$

with the GP operator H_{GP} given by the expression in the square brackets on the right hand side of Eq. (2.35). Here, γ is the dissipation strength and η a Gaussian white noise term defined by its second moment $\langle \eta(z,t)\eta(z',t')\rangle = 2\hbar\gamma k_{\mathrm{B}} T\, \delta(z - z')\delta(t - t')$. A thermal state is obtained by numerically propagating a seed state (e.g. an empty state or a state with a homogeneous density distribution) with Eq. (2.39) until convergence is reached. In this process γ defines the speed of the interaction with the incoherent thermal background and can be chosen arbitrarily as we are only interested in the final state and not the time evolution of the condensation dynamics. For coupled condensates two separate states $\psi_{1,2}$ are propagated in parallel with an additional $-\hbar J \psi_2(z)$ term added to the GP operator of the first condensate $H_{GP}[\psi_1]$, and vice versa.[12] The noise terms $\eta_{1,2}$ are independent for the two systems. The GPE simulations presented in Chap. 5 starting from thermal initial states were performed by Sebastian Erne [50].

References

1. Giamarchi T (2004) Quantum physics in one dimension. Clarendon Press, Oxford
2. Pitaevskii L, Stringari S (2003) Bose-Einstein condensation. Clarendon Press, Oxford
3. Mermin ND, Wagner H (1966) Absence of ferromagnetism or antiferromagnetism in one- or two-dimensional isotropic Heisenberg models. Phys Rev Lett 17:1133–1136
4. Hohenberg PC (1967) Existence of long-range order in one and two dimensions. Phys Rev 158:383–386
5. Kinoshita T, Wenger T, Weiss DS (2006) A quantum Newton's cradle. Nature 440:900–903
6. Caux J-S, Mossel J (2011) Remarks on the notion of quantum integrability. J Stat Mech Theory Exp P02023
7. Bloch I (2005) Ultracold quantum gases in optical lattices. Nat Phys 1:23–30
8. Kinoshita T (2004) Observation of a one-dimensional Tonks-Girardeau gas. Science 305:1125–1128
9. Paredes B et al (2004) Tonks-Girardeau gas of ultracold atoms in an optical lattice. Nature 429:277–281
10. Haller E et al (2009) Realization of an Excited, Strongly Correlated Quantum Gas Phase. Science 325:1224–1227
11. Reichel J, Vuletić V (eds) (2011) Atom chips. Wiley-VCH, Weinheim, Germany
12. Estève J et al (2006) Observations of density fluctuations in an elongated bose gas: ideal gas and quasicondensate regimes. Phys Rev Lett 96:130403
13. Hofferberth S, Lesanovsky I, Fischer B, Schumm T, Schmiedmayer J (2007) Non-equilibrium coherence dynamics in one-dimensional Bose gases. Nature 449:324–327
14. Pethick CJ, Smith H (2002) Bose-Einstein condensation in dilute gases. Cambridge University Press
15. van Kempen EGM, Kokkelmans SJJMF, Heinzen DJ, Verhaar BJ (2002) Interisotope determination of ultracold rubidium interactions from three high-precision experiments. Phys Rev Lett 88:093201
16. Lieb EH, Liniger W (1963) Exact analysis of an interacting bose gas. I. The general solution and the ground state. Phys Rev 130:1605–1616
17. Lieb EH (1963) Exact analysis of an interacting bose gas. II. The excitation spectrum. Phys Rev 130:1616–1624
18. Girardeau M (1960) Relationship between systems of impenetrable bosons and fermions in one dimension. J Math Phys 1:516–523
19. Yang CN, Yang CP (1969) Thermodynamics of a one-dimensional system of bosons with repulsive delta-function interaction. J Math Phys 10:1115–1122
20. Bouchoule I, Van Druten NJ, Westbrook CI (2011) Atom chips and one-dimensional bose gases. In: Reichel J, Vuletić V (eds) Atom Chips, Chap. 11. Wiley-VCH, Weinheim, Germany, pp 331–363
21. Stimming H-P, Mauser NJ, Schmiedmayer J, Mazets IE (2010) Fluctuations and stochastic processes in one-dimensional many-body quantum systems. Phys Rev Lett 105:015301
22. Mora C, Castin Y (2003) Extension of Bogoliubov theory to quasicondensates. Phys Rev A 67:053615

[12]In order to avoid spurious soliton excitations in the relative degrees of freedom of two coupled gases the seed state needs to have a finite occupation.

23. Baym G, Pethick CJ (1996) Ground-state properties of magnetically trapped bose-condensed rubidium gas. Phys Rev Lett 76:6–9
24. Tomonaga S-I (1950) Remarks on Bloch's method of sound waves applied to many-fermion problems. Prog Theor Phys 5:544–569
25. Luttinger JM (1963) An exactly soluble model of a many-fermion system. J Math Phys 4:1154–1162
26. Mattis DC, Lieb EH (1965) Exact solution of a many-fermion system and its associated boson field. J Math Phys 6:304–312
27. Haldane FDM (1981) Effective harmonic-fluid approach to low-energy properties of one-dimensional quantum fluids. Phys Rev Lett 47:1840–1843
28. Cazalilla MA (2004) Bosonizing one-dimensional cold atomic gases. J Phys B Atomic Mol Opt Phys 37:S1–S47
29. Petrov DS, Shlyapnikov GV, Walraven JTM (2000) Regimes of quantum degeneracy in trapped 1D gases. Phys Rev Lett 85:3745–3749
30. Dalton B, Ghanbari S (2012) Two mode theory of Bose-Einstein condensates: interferometry and the Josephson model. J Mod Opt 59:287–353
31. Langen T, Schweigler T, Demler E, Schmiedmayer J (2018) Double light-cone dynamics establish thermal states in integrable 1D Bose gases. New J Phys 20:023034
32. Schweigler T et al (2017) Experimental characterization of a quantum many-body system via higher-order correlations. Nature 545:323–326
33. Cuevas-Maraver J, Kevrekidis PG, Williams F (eds) (2014) The sine-Gordon model and its applications. Springer
34. Mazets IE, Schmiedmayer J (2008) Dephasing in two decoupled one-dimensional Bose-Einstein condensates and the subexponential decay of the interwell coherence. Eur Phys J B 68:335–339
35. Stimming H-P, Mauser NJ, Schmiedmayer J, Mazets IE (2011) Dephasing in coherently split quasicondensates. Phys Rev A 83:023618
36. Huber S, Buchhold M, Schmiedmayer J, Diehl S (2018) Thermalization dynamics of two correlated bosonic quantum wires after a split. Phys Rev A 97:043611
37. Görlitz A, et al. (2001) Realization of Bose-Einstein condensates in lower dimensions. Phys Rev Lett 87:130402
38. Olshanii M (1998) Atomic scattering in presence of an external confinement and a gas of impenetrable bosons. Phys Rev Lett 81:938–941
39. Haller E et al (2010) Confinement-induced resonances in low-dimensional quantum systems. Phys Rev Lett 104:153203
40. Salasnich L, Parola A, Reatto L (2002) Effective wave equations for the dynamics of cigar-shaped and disk-shaped Bose condensates. Phys Rev A 65:043614
41. Salasnich L, Parola A, Reatto L (2004) Dimensional reduction in Bose-Einstein-condensed alkali-metal vapors. Phys Rev A 69:045601
42. Menotti C, Stringari S (2002) Collective oscillations of a one-dimensional trapped Bose-Einstein gas. Phys Rev A 66:043610
43. Mazets IE, Schmiedmayer J (2010) Thermalization in a quasi-one-dimensional ultracold bosonic gas. New J Phys 12:055023
44. Mazets IE, Atominstitut, TU Wien, Austria, igor.mazets@tuwien.ac.at (Personal communication)
45. Mazets IE, Schumm T, Schmiedmayer J (2008) Breakdown of integrability in a quasi-1D ultracold bosonic gas. Phys Rev Lett 100:210403
46. Schweigler T (2019) Correlations and dynamics of tunnel-coupled one-dimensional Bose gases. Ph.D. thesis, TU Vienna
47. Bao W, Jaksch D, Markowich PA (2003) Numerical solution of the Gross-Pitaevskii equation for Bose-Einstein condensation. J Comput Phys 187:318–342
48. Rohringer W (2014) Dynamics of one-dimensional bose gases in time-dependent traps. Ph.D. thesis, TU Vienna
49. Gardiner CW, Anglin JR, Fudge TIA (2002) The stochastic Gross-Pitaevskii equation. J Phys B Atomic Mol Opt Phys 35:1555–1582
50. Erne S (2018) Far-from-equilibrium quantum many-body systems from universal dynamics to statistical mechanics. Ph.D. thesis, Ruperto-Carola University of Heidelberg

Chapter 3
Experimental Setup and Probing

This chapter serves as an introduction to the experimental setup. It discusses the tools and techniques developed to create and probe the 1d Bose gases described in Chap. 2. First, Sect. 3.1 reviews basic experimental techniques until Sect. 3.2 describes the experimental setup and cycle. Both of these discussion are kept brief as there exists a vast body of literature and theses on these topics. Finally, in Sect. 3.3, the techniques employed to probe the 1d Bose gases are discussed in detail, from time-of-flight expansion to methods of extracting information from absorption images.

3.1 Trapping and Cooling Neutral Atoms

Over the last decades various techniques have been developed to trap, store and cool neutral atoms. In the following, the ones most relevant to our experimental setup will be discussed briefly. The focus of this discussion lies on topics which are either not ubiquitously covered in the literature or are of particular importance for the results presented in this thesis.

First, a review of magnetic trapping, mainly covering wire traps and radio-frequency (rf) dressed state potentials is given. The wire layouts and field configurations employed in our setup are at the center of this discussion. In the following, the foundations of optical dipole potentials are discussed, since their introduction to the experiment was part of this thesis (see Sect. 3.2.3). Finally, evaporative cooling is reviewed shortly as its comprehension poses a prerequisite to understand the novelty of the results presented in Chap. 4. For discussions of the magneto-optical trap (MOT), laser cooling and optical pumping the reader is referred to Refs. [1–3].

B. Rauer, *Non-Equilibrium Dynamics Beyond Dephasing*,
Springer Theses, https://doi.org/10.1007/978-3-030-18236-6_3

3.1.1 Magnetic Trapping

Magnetic traps for neutral atoms are based on the local energy shift an atom experiences in a inhomogeneous magnetic field $\mathbf{B}(\mathbf{r})$. For an atom with a magnetic dipole moment $\boldsymbol{\mu}_\mathrm{m}$ this is given by $-\boldsymbol{\mu}_\mathrm{m} \cdot \mathbf{B}$. If this shift is small enough such that the magnitude of the total angular momentum $\mathbf{F}$ is still a good quantum number the resulting potential can be written as

$$U_\mathrm{mag}(\mathbf{r}) = g_F\, \mu_\mathrm{B}\, \mathbf{F} \cdot \mathbf{B}(\mathbf{r}) = m_F\, g_F\, \mu_\mathrm{B} |\mathbf{B}(\mathbf{r})| \,, \tag{3.1}$$

with g_F being the Landé factor and μ_B the Bohr magneton. For the last equality we assumed the atomic angular momentum always to be aligned with the local magnetic field, m_F being the quantum number of the $\mathbf{F}$ component along this direction. This assumption is justified as long as, in the frame of a moving atom, the direction of the magnetic field changes slower than the Larmor frequency $\omega_\mathrm{L} = m_F\, g_F\, \mu_\mathrm{B}|\mathbf{B}|/\hbar$, such that $\partial_t \mathbf{B}/|\mathbf{B}| \ll \omega_\mathrm{L}$. Is this the case, then the orientation of the magnetic moment can adiabatically follow the local magnetic field.

Since Maxwell's equations do not allow for free space maxima of the magnetic field, only atoms in low-field seeking states with $m_F\, g_F > 0$ can be trapped magnetically [4]. For ^{87}Rb used in our setup, the two hyperfine states $F = 1, 2$ of the $5^2S_{1/2}$ ground state have the Landé factors $g_F \simeq -1/2$ and $g_F \simeq 1/2$, respectively [5]. This means that the state $|F = 1, m_F = -1\rangle$ as well as $|2, 1\rangle$ and $|2, 2\rangle$ can be trapped magnetically, the latter of which is used in our setup. This state selectivity poses an advantage for magnetic traps when employing evaporative cooling as rf induced spin flips to untrapped states can be used to remove atoms at a defined energy (see Sect. 3.1.5). However, it also severely restricts the accessible internal state dynamics.

Note that the above conditions for the Larmor frequency is violated in regions of zero or low magnetic fields, leading to so-called Majorana spin flips to possibly untrapped states resulting in loss [6]. In hot clouds, it is less probable for atoms to be found at the minimum of the trap and these losses therefore do not pose a problem.

3.1.2 Wire Traps

In our setup, magnetic trapping potentials are created by current carrying wires and homogeneous bias field. These wires are either solid macroscopic copper structures or microfabricted films on a chip surface, however, the basic principles of the realized trapping potentials are the same. Sending the current I_w through an infinitely thin wire creates a magnetic field pointing around the wire given by

$$\mathbf{B}_\mathrm{w}(r) = \frac{\mu_0 I_\mathrm{w}}{2\pi r} \hat{\mathbf{e}}_\phi \,,$$

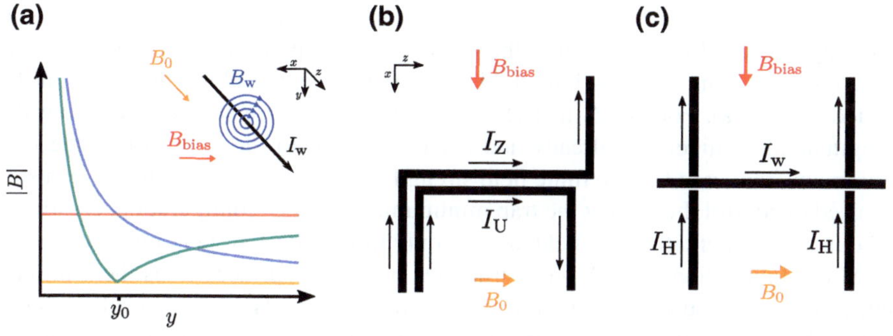

Fig. 3.1 **Wire traps. a** Magnetic field magnitude of a wire guide trap (green) created at a distance y_0 below the wire. The trap arises from adding the magnetic field of an ideal wire B_w (blue) and a homogeneous bias field B_{bias} (red) oriented perpendicular to the wire. An additional bias field B_0 (yellow) aligned along z lifts the minimum of the trap away from zero magnetic field. The inset shows a schematics of the field orientations and axis conventions. **b** Planar schematics of two single wire trap configurations in the shape of a "Z" and the shape of a "U". **c** Schematics of a multi-wire configuration in the shape of an "H". Figures inspired by Ref. [7]

where μ_0 is the vacuum permeability and $\hat{\mathbf{e}}_\phi$ the angular unit vector. The radial distance from the wire is given by $r^2 = x^2 + y^2$ in our choice of coordinates (Fig. 3.4a inset). Adding a homogeneous bias field $\mathbf{B}_{bias} = B_{bias}\hat{\mathbf{e}}_x$ to this configuration, oriented perpendicular to the wire, leads to a cancellation of the fields at a distance defined by the wire current and the magnitude of the bias field given by

$$y_0 = \frac{\mu_0 I_w}{2\pi B_{bias}} .$$

This creates a trap with a quadrupole field configuration that harmonically confines atoms in the xy-plane (Fig. 3.1a), however, they can still move freely along the wire.

To realize a 3d trap, confining the atoms also along z, additional wires are needed. One of the simplest configuration to achieve this is by bending the two ends of the wire into a "Z" shape (Fig. 3.1b). The fields created by these perpendicular leads, which are typically further away from the atoms, result in a weak harmonic confinement along the z direction. Together with the central section and the bias field, this adds to a Ioffe-Pritchard trap configuration [6]. Bending both leads in the same direction, giving the wire a "U" shape, creates a 3d quadrupole configuration (Fig. 3.1b).

Using multiple wires, one gains more flexibility. For example, a straight trapping wire augmented with two perpendicular wires responsible for the longitudinal confinement along z realizes a Ioffe-Pritchard field configuration as well (Fig. 3.1c). The advantage of this "H" configuration over the Z-trap discussed above is that the longitudinal confinement can be changed independently of the transverse one. Additionally, the H-configuration is completely symmetric while the Z-trap is slightly tilted in the xz-plane with respect to the central wire segment.

For the wire guide realized by a single straight wire and a transverse bias field the magnetic field goes to zero in the center of the trap. To avoid Majorana losses an additional homogeneous bias field $\mathbf{B}_0 = B_0\hat{\mathbf{e}}_z$ oriented along the wire can be added (Fig. 3.1a). For the Z- and H-configuration, the wire segments creating the longitudinal confinement already add a z-component to the field, lifting the trap minimum from zero field to finite field strengths. Yet, the bias field along z can still be used to control the field at the trap minimum. For the U-configuration, creating a 3d quadrupole field, the bias field along z only moves the zero field minimum.

Micro-fabrication of such wire structures on a chip surfaces turns them into an integrated and robust platforms for cold atom experiments. These so-called 'atom chips' allow for the trapping of atoms close to the current carrying structures, realizing strong field gradients without resorting to excessive currents. Especially, the quasi-1d confinements discussed in Sect. 2.2 can be realized effectively due to the elongated trap geometry. For reviews on magnetic micro-traps and atom chips covering the fabrication process, real wire effects and the different configurations employed see Refs. [8–10].

3.1.3 Radio-Frequency Dressed State Potentials

Another advantage of atom chips is that they are particularly well suited to implement rf dressed state potentials. These potentials arise when strongly coupling the atomic Zeeman states through an oscillating magnetic field, resulting in a new set of dressed eigenstates. Depending on the geometry of the coupling, the composition of these eigenstates can vary in space. Just as for static magnetic traps, a moving atom can adiabatically follow these changes and always remain in the same state. This spatial dependence of the eigenstate structure enables confinement geometries beyond what is possible with static magnetic fields. In atom chips, dedicated wires carrying rf currents can, due to their proximity, deliver strong oscillating fields at the position of the atoms. Furthermore, the superposition of phase-shifted fields from multiple wires allows for a precise control of the rf field polarization. In the following, we will briefly discuss rf dressed state potentials and give an example relevant to our setup. For a more detailed discussion see Ref. [11] and for a pedagogical introduction to rf dressed state potentials Ref. [12].

The interaction of an atom subject to a static magnetic field $\mathbf{B}_{\text{stat}}(\mathbf{r})$ and a rapidly oscillating field with an amplitude $\mathbf{B}_{\text{rf}}(\mathbf{r})$ is described by the Hamiltonian[1]

$$H_{\text{mag}} = g_F\,\mu_{\text{B}}\,\mathbf{F}\cdot\left[\mathbf{B}_{\text{stat}}(\mathbf{r}) + \mathbf{B}_{\text{rf}}(\mathbf{r})\cos\left(\omega_{\text{rf}}\,t\right)\right], \tag{3.2}$$

with ω_{rf} being the frequency of the rf field oscillation. Assuming that the spin adiabatically follows the static field direction we can perform a local rotation of coordinates

[1] As for Eq. (3.1), this expression is valid as long as the fields are small enough such that the magnitude of $\mathbf{F}$ remains a good quantum number.

to a frame where the spin is always aligned in z-direction. From this we immediately see that only components of the rf field oriented perpendicular to the local static field, $\mathbf{B}_{\mathrm{rf},\perp}(\mathbf{r})$, are coupling the spin states. Further, going to a frame co-rotating with the rf field and neglecting the term oscillating at $2\omega_{\mathrm{rf}}$ in the rotating wave approximation, we can diagonalize the local Hamiltonian ending up with the adiabatic potential

$$U_{\mathrm{rf}}(\mathbf{r}) = m'_F\, g_F\, \mu_{\mathrm{B}} \sqrt{\left(|\mathbf{B}_{\mathrm{stat}}(\mathbf{r})| - \frac{\hbar\omega_{\mathrm{rf}}}{|g_F|\mu_{\mathrm{B}}}\right)^2 + \left(\frac{|\mathbf{B}_{\mathrm{rf},\perp}(\mathbf{r})|}{2}\right)^2}$$
$$= m'_F\, \hbar \sqrt{\Delta(\mathbf{r})^2 + \Omega(\mathbf{r})^2}\,.$$

Here, m'_F is the quantum number of the new dressed basis. The geometry of the potential is defined by the local rf detuning $\Delta(\mathbf{r})$ and the Rabi frequency $\Omega(\mathbf{r})$. The latter is proportional to the magnitude of $\mathbf{B}_{\mathrm{rf},\perp}(\mathbf{r})$ while the first is given by the local energy difference between the Zeeman splitting and ω_{rf}.

On an atom chip, placing two wires carrying rf currents adjacent to a straight trapping wire of the kind discussed in Fig. 3.1a, we can create a nearly homogeneous rf field at the location of the trap minimum (Fig. 3.2a). The orientation of this field can be controlled by the relative phase of the currents in the two wires. Setting the phase to π results in a linearly polarized rf field oriented perpendicular to the chip surface. Together with the quadrupole field orientation of the static trap in the xy-plane, this leads to a horizontal double well potential (Fig. 3.2). For negative detunings, where ω_{rf} is larger than the Zeeman splitting in the center of the trap, there is always a double well forming. In the case of positive detuning, the double well forms at a critical rf field strength, when $\Omega(\mathbf{r})$ overpowers $\Delta(\mathbf{r})$. The latter is used in our setup. In this configuration, when the current in the rf wires is ramped up slowly, the atoms seamlessly cross from the bare states to the dressed eigenstates such that the trap smoothly transforms from a single well to a double well.

For more information on the different trap geometries that can be realized with rf dressed state potentials see Refs. [11, 13–15]. For an analysis of effects beyond the rotating wave approximation see Ref. [16].

3.1.4 *Optical Dipole Potentials*

Another way to confine atoms to a defined region of space are optical dipole potentials. They utilizing the electric dipole interaction of an atom with a light field that is far detuned from an atomic transition. The following section discusses the topic in broad strokes. For a more detailed treatment the reader is directed to Ref. [17].

An intuitive framework to understand optical dipole potentials is the dressed state picture, already employed in Sect. 3.1.3, and time-independent perturbation theory. As the electric dipole moment μ_{e} is induced by the field E experienced by the atom, it is necessary to go to second order in perturbation theory for the leading energy correction. Assuming a two-level atom this correction is given by

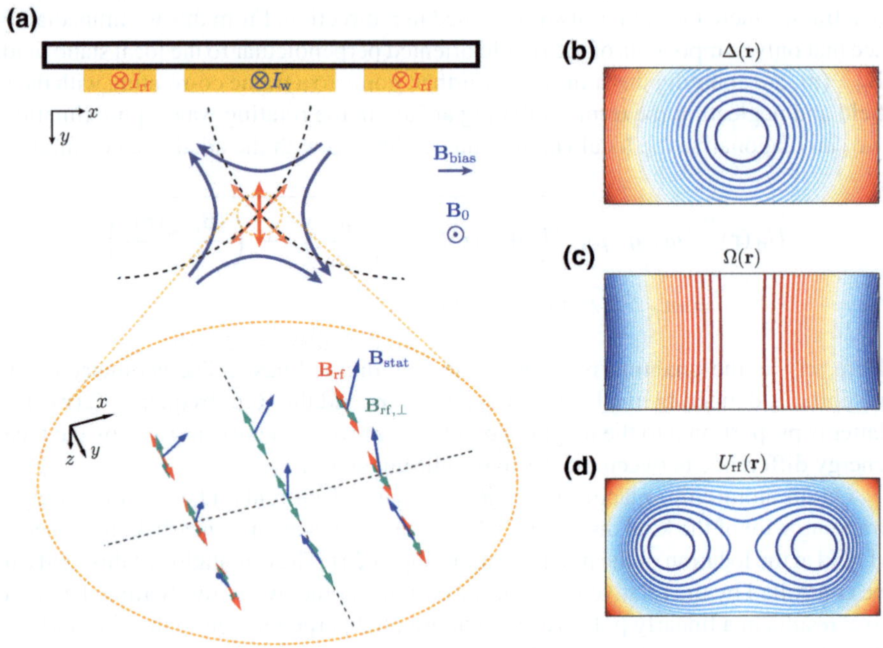

Fig. 3.2 Rf dressed state potentials on a chip. a Schematics of magnetic field orientations creating the dressed state potentials in our setup. The static current I_w through the main trapping wire, together with the transverse bias field $\mathbf{B}_{bias}$ and the longitudinal field $\mathbf{B}_0$, creates the quadrupole trapping field $\mathbf{B}_{stat}$ (blue). The alternating currents I_{rf} in the two rf wires give rise to oscillating fields that add to $\mathbf{B}_{rf}$, oriented along y perpendicular to the chip surface (red). In the 3d schematics of the field orientations in the trap center (yellow circle) we see that the rf component perpendicular to the static field direction, $\mathbf{B}_{rf,\perp}$ (green), decreases when leaving the trap center in x direction but remains constant when moving along y. The dashed black lines in the inset act as guides to the eye, tracing the orientation of the x and y axis. **b** Contour plot showing the spatial detuning of the rf from the local Zeeman energy splitting, reflecting the bare harmonic trap. The color map goes from blue (small values) to red (large values). **c** Contour plot of the coupling strength proportional to $\mathbf{B}_{rf,\perp}$. The spatial structure arises from the relative field orientations observed in (**a**). **d** Contour plot of the resulting adiabatic potential for a positive rf detuning

$$\delta\varepsilon = \frac{|\langle e|V|g\rangle|^2}{\varepsilon_g - \varepsilon_e}, \tag{3.3}$$

where $V = -\mu_e E$ is the interaction term with the electric field and $\varepsilon_{g,e}$ are the energies of the ground state $|g\rangle$ and the excited state $|e\rangle$, respectively. Considering the full system of light and atom in the states $|g\rangle$ and $|e\rangle$, one needs to take the electric field into account. If the atom is in the ground state, it has no internal energy. The total energy of the combined state with the light field is then given by $\varepsilon_g = n\hbar\omega$, with ω being the laser frequency and n the mode occupation. To get to the excited state, the atom needs to absorb a photon, such that $\varepsilon_e = \hbar\omega_0 + (n-1)\hbar\omega$, where $\hbar\omega_0$ is the energy of the atomic transition. The energy difference in the denominator of Eq.

(3.3) is therefore given by $\varepsilon_g - \varepsilon_e = \hbar(\omega - \omega_0) = \hbar\Delta$. Using the relation between the dipole matrix element and the line width

$$\Gamma = \frac{\omega_0^2}{3\pi\epsilon_0\hbar c^3}|\langle e|\mu_e|g\rangle|^2,$$

we therefore obtain for the energy shift

$$\delta\varepsilon = \frac{|\langle e|\mu_e|g\rangle|^2}{\hbar\Delta}|E|^2 = \frac{3\pi c^2}{2\omega_0^3}\frac{\Gamma}{\Delta}I. \tag{3.4}$$

For the last equality we averaged over the rapidly oscillating field and inserted $I = 2\epsilon_0 c|E|^2$ for the intensity of the laser light. The energy obtained in Eq. (3.4) gives the light shift of the atomic ground state. The excited state experiences the opposite shift as the denominator in Eq. (3.3) changes sign. However, as dipole potentials aim to minimize scattering and therefore work with low saturation, the atom predominantly resides in the ground state, such that Eq. (3.4) can be seen as the potential experienced by the atom. As the detuning Δ defines the sign of this potential it determines whether it is repelling or attractive. For blue detuned light ($\Delta > 0$) the atoms are repelled from regions of high intensity while for red detuned light ($\Delta < 0$) they are attracted.

For real atoms with many transitions one needs to sum Eq. (3.3) over all contributing lines, taking the different transition matrix elements into account. In the case of alkali atoms the fine structure splitting of the D line leads, for linearly polarized light, to the dipole potential[2]

$$U_{\mathrm{od}}(\mathrm{r}) = \frac{\pi c^2\Gamma}{2\omega_0^3}\left(\frac{1}{\Delta_{D_1}} + \frac{2}{\Delta_{D_2}}\right) I(\mathrm{r}), \tag{3.5}$$

with Δ_{D_1} and Δ_{D_2} being the individual detunings to the D_1 ($^2S_{1/2} \rightarrow {}^2P_{1/2}$) and the D_2 ($^2S_{1/2} \rightarrow {}^2P_{3/2}$) line. For circularly polarized light not discussed here, an explicit dependence on the hyperfine state and the magnetic substructure enters. Similar to Eq. (3.5), and under the same assumptions, the scattering rate of dipole light photons by the atom can be shown to be

$$\Gamma_{\mathrm{sc}}(\mathrm{r}) = \frac{\pi c^2\Gamma^2}{2\hbar\omega_0^3}\left(\frac{1}{\Delta_{D_1}^2} + \frac{2}{\Delta_{D_2}^2}\right) I(\mathrm{r}).$$

As $\Gamma_{\mathrm{sc}}/U_{\mathrm{od}}$ scales with Δ^{-1}, it is beneficial to go to large detunings and high laser intensities in order to avoid scattering. Also, traps created from blue detuned light are advantageous to obtain minimal scattering as there atoms typically reside in regions of low intensity.

---

[2]Assuming that all detunings are much larger than the excited state hyperfine splitting.

In comparison to magnetic traps optical dipole traps are typically much shallower. If scattering is meant to be negligible, large detunings and laser powers in the range of Watts are needed to achieve potentials with millikelvin depths. Further, evaporative cooling does not work as straightforward as for magnetic traps (see Sect. 3.1.5). To selectively release the most energetic particles, the trap needs to be ramped down which can change the shape of the confinement.

An advantageous feature of dipole traps is that, under the right conditions, different internal states experience the same potential. This opens the possibility to utilize internal ground state dynamics to a degree not accessible with magnetic traps. Another striking advantage of optical dipole traps is their flexibility. Laser light can easily be shaped, superposed and interfered to create complex trapping geometries ranging from optical lattices [18] or flat bottom box potentials [19–21] up to fully arbitrary 3d arrays of trapping sites [22]. Section 3.2.3 will discuss how we exploit this flexibility to create a 1d box trap and in Sect. 6.1 a new light shaping setup will be discussed that enables an arbitrary control over the longitudinal potential of our 1d system.

3.1.5 Evaporative Cooling

Apart from trapping, efficient cooling techniques are imperative for the preparation of cold atomic gases. Beyond the limits of laser cooling, cooling by evaporation is the prime tool to achieve ultracold temperatures and reach degeneracy. First devised for spin-polarized hydrogen [23], its implementation for cold gases of alkali atoms represented the last step towards the experimental realization of BEC [24, 25].

Evaporative cooling works through an energy selective removal of particles and consecutive rethermalization. The most energetic particles are expelled from the trap, reducing the average energy per particle such that the gas equilibrates to a lower temperature (Fig. 3.3). For this mechanism to work, the gas needs to be sufficiently ergodic, meaning that an atom with a certain energy can sample the entire phase space corresponding to that energy, and thermalization needs to proceed fast compared to the evaporation rate. The cooling efficiency depends on the density of states and therefore on the trap geometry. Under realistic conditions, the cooling also needs to be faster than any loss or heating processes limiting the lifetime of the gas. For a semi-classical treatment of evaporative cooling see Refs. [26, 27]. Chapter 4 will discuss how this process translates to a degenerate 1d Bose gas where thermalization is highly suppressed, which should render the cooling ineffective.

For magnetically trapped samples, energy selective outcoupling can efficiently be implemented by rf or microwave induced spin-flips to untrapped states. These spin-flips are driven by oscillating magnetic fields through the interaction term given by Eq. (3.2). Setting the rf frequency to the desired outcoupling energy above the bottom of the trap opens the potential and lets particles that reach this point transition to

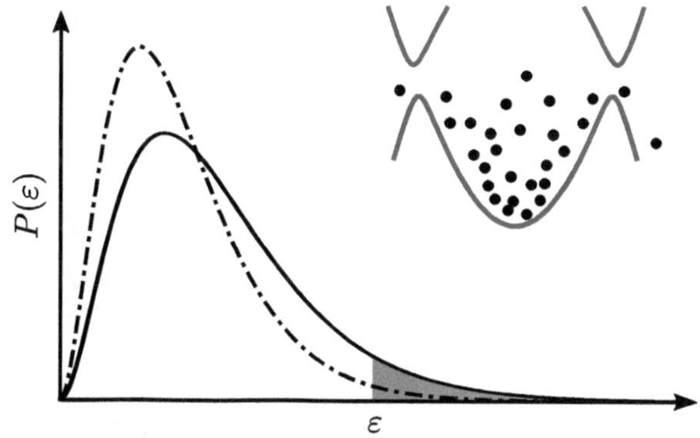

Fig. 3.3 Evaporative cooling. Schematic depiction of standard evaporative cooling. Shown is the Maxwell-Boltzmann distribution of particles with the energy ε in a harmonic trap. Cutting the most energetic particles (shaded area) from a hot cloud (solid line) results after rethermalization in a colder distribution (dash-dotted line). The inset depicts cooling through rf outcoupling from an harmonic trap

untrapped states (Fig. 3.3 inset). For ^{87}Rb atoms in $|F = 2, m_F = 2\rangle$ either all states within the $F = 2$ manifold can be coupled via rf fields or microwave fields can be used to drive the transition to $|F = 1, m_F = 1\rangle$.[3]

3.2 Experimental Setup

Initially designed to investigate Bose-Fermi mixture of ^{40}K and ^{87}Rb, building of the experimental apparatus started about 10 years prior to the measurements presented in this thesis. The goal was to combine the low-dimensional trapping geometry of an atom chip with the physics of Bose-Fermi mixtures. Main features of the setup were envisioned to be versatile species-selective rf dressed state potentials, allowing for the implementation of 2d tube-like traps [11], and a high resolution absorption imaging system working in reflection from the chip surface (see Sect. 3.3.2) [28]. The design and conception of the setup is described in detail in the theses of the first generation of students [15, 29, 30]. The theses of the successive students cover the final construction and improvements to the setup [31–36].

In the following, a brief overview of the setup and the procedure to create a degenerate 1d Bose gas is given. This intends only to give a broad reference for the measurements discussed in Chaps. 4 and 5. For details the reader is directed to the comprehensive body of theses. Only the newly implemented dipole trap setup will be discussed in full detail.

[3]The ground state hyperfine splitting in ^{87}Rb is 6.8 GHz [5] while the for typical parameters in our setup the Zeeman state splitting in the center of the undressed trap is ~390 kHz (see Sect. 3.2.2).

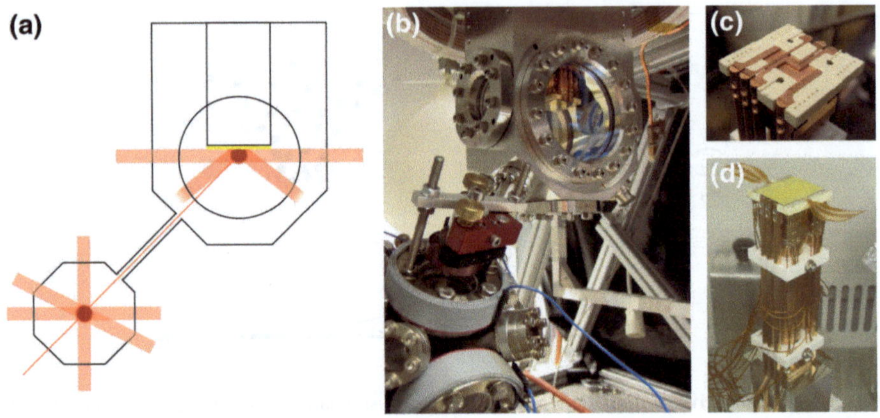

Fig. 3.4 **Experimental setup.** **a** Schematics of the experimental setup showing the source chamber (lower left) and the science chamber containing the atom chip (upper right). The MOT beams are depicted as thick red lines while the beam pushing the atoms through the differential pumping stage is shown as a thin red line. **b** Picture of the two vacuum chambers. **c** Chip mounting containing the macroscopic copper wire structures. **d** Mounting rod with the chip on top

3.2.1 Overview

The setup is built around two vacuum chambers (Fig. 3.4a and b). A science chamber with good optical access houses the atom chip while an axillary source chamber is equipped with atom dispensers for Potassium and Rubidium. The two chambers are linked by a differential pumping stage allowing the science chamber to maintain better vacuum.[4] In the source chamber, atoms are gathered from the background vapor in a standard six-beam 3d MOT. While this source-MOT is loading, the trapped atoms are continuously pumped through the differential pumping stage and into the science chamber by a resonant laser beam (Fig. 3.4a). In the science chamber, they are collected by a second 3d MOT which is formed in reflection just below the atom chip [37, 38]. The chip is mounted upside-down on a rod connected to the ceiling of the vacuum chamber (Fig. 3.4d). Inside this rod, just above the downwards facing chip surface, macroscopic copper wire-structures are placed (Fig. 3.4c). Sending current though a U-shaped wire, in combination with a homogeneous external bias field, creates the quadrupole field for the second MOT (see Sect. 3.1.2).

After a 10 s collection phase the second MOT is turned off for a ~10 ms long stage of polarization gradient cooling in the σ^+-σ^- configuration [3]. Following that, the atoms are pumped into the $|F = 2, m_F = 2\rangle$ state for optimal load-over to an intermediate magnetic trap. This magnetic trap is formed by current flowing in two Z-shaped copper wires above the chip (Fig. 3.4c) and homogeneous external fields (see Sect. 3.1.2). It provides a much deeper and wider confinement than the final chip trap and is used for compression and an initial evaporative cooling stage. After

[4] 10^{-10} mbar, pumped by an ion pump (Varian 150 l/s Star-Cell) and a titanium sublimation pump.

about 6 s of evaporation we obtain a thermal cloud of $\sim 2 \cdot 10^6$ atoms at $1-2\,\mu$K. By ramping down the current in the Z-shaped wires while sending current through the main trapping wire of the atom chip, the atoms are finally loaded into the chip wire trap. There, a second stage of evaporation cools the gas to degeneracy.

For both evaporation stages, rf fields energy-selectively couple atoms to untrapped Zeeman states $m_F \leqslant 0$. An arbitrary waveform generator[5] programmed to output sine waves of adjustable frequency is connected to the U-shaped wire below the chip, bringing the rf current close to the position of the trap. During cooling, combinations of linear and exponential frequency ramps, starting at 15 MHz and ending between 390 and 520 kHz, continuously eject the most energetic particles from the trap [36]. Alternatively, a microwave source[6] connected to an antenna outside the chamber can be used to drive transitions to untrapped states in the $F = 1$ manifold [35].

The laser light for cooling, pumping and imaging, tuned close to the $F = 2 \rightarrow F' = 3$ transition of the ^{87}Rb D_2 line, is obtained from a single frequency Ti:sapphire laser.[7] Light locked to the $F = 1 \rightarrow F' = 2$ transition, used to pump atoms that accumulate in $F = 1$ back into resonance with the cooling and pumping lasers is generated by a DFB diode laser.[8]

3.2.2 *Atom Chip Trap*

The atom chip is a microfabricated assembly of gold wires structured onto a silicon wafer with a thin isolating layer of SiO_2 in between [15]. The wires are 10–100 μm wide and individually bonded to leads leading outside the chamber. The empty areas in between the wires are covered in gold to achieve maximum reflectivity. This is important for the reflective MOT discussed above and the transverse imaging system where the imaging light is reflected off the chip surface (Sect. 3.3.2). Although the chip is equipped with two independent main trapping wires both supplemented by two adjacent rf wires, and three pairs of U-shaped wires, only the ones shown in the schematics given in Fig. 3.5 are used in the measurements presented in this thesis. For a detailed layout of the wires see [32].

As discussed in Sect. 3.1.2 the strong transverse confinement is created by adding the field of a single main trapping wire with a homogeneous bias field. In our case this wire is 100 μm wide and carries a current of $I_w = 820$ mA. The bias field is typically 15.5 G, forming the trap minimum about 100 μm below the chip surface. For the longitudinal confinement, the U-wires shown in Fig. 3.5 approximate an H-configuration, with $I_U = 307$ mA and $I_U = 196$ mA flowing through the first and

[5]Tabor Electronics WW1071.

[6]A 6.81 GHz oscillator mixed with a sweepable $0-30$ MHz source (SRS DS345) or/and a sweepable $1-150$ MHz source (VFG150).

[7]Coherent MBR-110 pumped by a Coherent Verdi V18.

[8]Toptica LD-0780-0080-DFB-1 in a ThorLabs LDM21 mount.

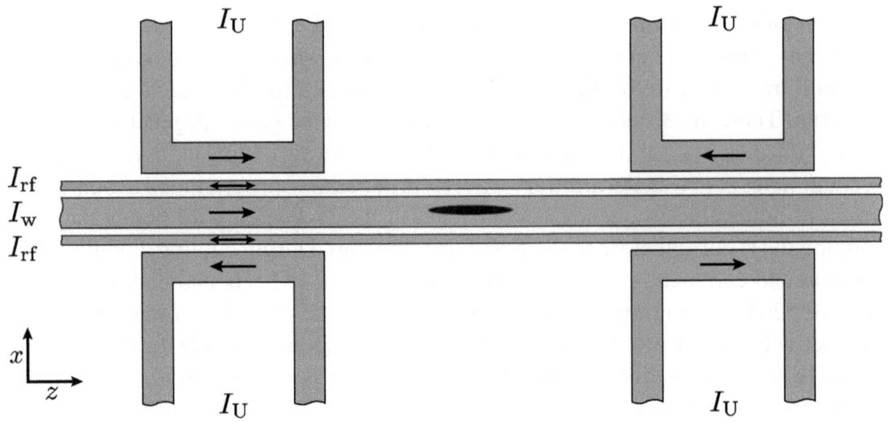

Fig. 3.5 Chip wire layout. Schematics of the chip wires with the arrows indicating the direction of current flow. The atom cloud is indicated by the black elongated ellipse. The depiction is not to scale and only the wires used in the measurements presented in this thesis are shown

second pair, respectively[9] (see Sect. 3.1.2). Additionally, a homogeneous bias field along the z-direction is applied to lift the frequency of the Zeeman energy splitting in the trap center to 390 kHz. The final confinement has typical trap frequencies of $\omega_\perp = 2\pi \cdot 2.1$ kHz in the two tightly confined transverse directions and $\omega_z = 2\pi \cdot 11$ Hz in the weakly confined longitudinal direction.

For the dressed trap discussed in Sect. 3.1.3, the necessary oscillating magnetic field is generated by applying rf currents to two 30 µm-wide wires next to the main trapping wire (Fig. 3.5). To achieve an oscillating field pointing along y, creating a double well orientated perpendicular to gravity along x, the currents are set to a relative phase shift of π (see Sect. 3.1.3). Typical currents amplitudes of $I_{rf} = 24$ mA at a frequency tuned ~30 kHz below the Zeeman splitting in the trap center lead to a fully decoupled double well for the standard static trap parameters discussed above. The trap frequencies in each of the wells are typically $\omega_\perp = 2\pi \cdot 1.45$ kHz and $\omega_z = 2\pi \cdot 7$ Hz. Lower currents create coupled double wells or deformed single wells.

A limiting factor for our setup are corrugations in the main trapping wire, typical for atom chips [39, 40]. They lead to imperfect current distributions in the wires and ultimately to unwanted spatial modulations of the magnetic potential. Moving the trap further away from the wires diminishes the contribution of short length scale modulations and allows to select regions along the wire where an unperturbed trap can be formed. However, this puts constraints on the realizable trap frequencies, such that the frequencies given above are the upper limit for $\omega_\perp$ and the lower limit for

[9] In Fig. 3.5, the first and second U-wire pairs correspond to the left and right U-wires. A difference in the currents flowing in these pairs allows to move the trap along the main trapping wire. For us, the optimal position is chosen by finding a region where the potential is sufficiently flat, least perturbed by wire corrugations.

ω_z at the chosen position along the wire. The dipole trap setup discussed in Sect. 6.1 is designed to smoothen the trap modulations, allowing for more flexibility of the magnetic trap position.

3.2.3 *Box Trap*

While being a robust and integrated platform for cold atoms experiments, atom chips lack the flexibility of optical dipole traps. For the chip design used in our setup a constraint particularly interesting to overcome is the limitation to harmonic confinements along the longitudinal axis of the trap. Especially desirable are flat bottom box traps allowing the preparation of samples with nearly homogeneous density profiles. Such confinements were previously realized for optically trapped bosonic systems in various dimensionalities [19–21]. Being closer to the condensed matter systems often inspiring cold atoms quantum simulations, such traps can enable the straight forward inference of bulk properties and render the local density approximation unnecessary. For 1d systems on atom chips, box-like confinements have been realized through local bends in the main trapping wire [41]. However, this approach is not particularly flexible as the properties of the trap are defined with the chip fabrication and any extensions to more complicated geometries require a new chip design. In our setup we therefore opt to augment the magnetic chip trap with a superposed optical dipole potential. This approach allows to maintain the advantages of the chip trap, such as rf dressed state potentials and effective evaporative cooling, while adding flexibility to the longitudinal confinement.

The optical potential is applied by shining light on the atoms from a direction transverse to the longitudinal axis of the trap. We use a blue detuned Gaussian beam[10] that is sent on an exchangeable wire mask target. The mask target is imaged onto the trap through the objective of the transverse imaging system (see Sect. 3.3.2). It cuts away the central part of the beam leaving two repelling, steep potential walls at its edges, thereby forming the box confinement (Fig. 3.6a). The length of the box is set by the width of the mask and the 4.6 fold demagnification of the imaging.[11] Its wall steepness is defined by the resolution of the mask imaging. Assuming that any point on the mask is imaged as a Gaussian intensity spot of width σ_{box} we can convolve an ideal box potential of length L and height U_0 with the effective point spread and obtain

$$U_{\text{box}}(z) = \frac{U_0}{2}\left[\operatorname{erfc}\left(\frac{z + L/2}{\sigma_{\text{box}}\sqrt{2}}\right) + \operatorname{erfc}\left(\frac{-z + L/2}{\sigma_{\text{box}}\sqrt{2}}\right)\right],$$

[10] $\lambda = 767$ nm, Toptica DLX 110.

[11] The optical setup was previously used with resonant light. It was designed to selectively pump parts of the cloud into a dark state for tomographic imaging. More information can be found in [31] and the supplementary materials of [42].

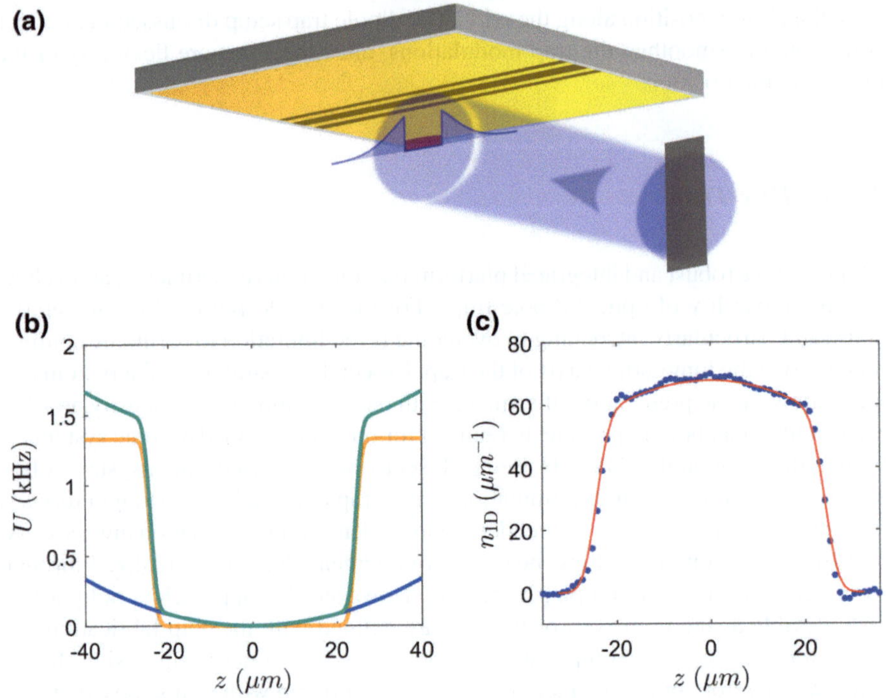

Fig. 3.6 **Box trap. a** Schematics of the box trap setup. Blue detuned laser light is sent on a mask that is imaged onto the atoms. Light is blocked from the center of the beam forming a box with steep potential walls. **b** Box potential (yellow), harmonic potential (blue) and their combination (green) for typical experimental parameters $U_0 = 1.3\,\text{kHz}$, $\sigma_{\text{box}} = 1.2\,\mu\text{m}$, $L = 49\,\mu\text{m}$ and $\omega_z = 2\pi \cdot 7\,\text{Hz}$. **c** Measured quasi-in situ density profile after 2 ms of expansion. The red line corresponds to the density profile of the broadned GPE ground state in the combined potential shown in (**b**). It is obtained from an imaginary time evolution (see Sect. 2.3.2) and takes the finite imaging resolution into account

with erfc(z) being the complementary error function. To obtain the full longitudinal potential experienced by the atoms, the harmonic confinement of the magnetic trap has to be added, leading to a slight curvature of the box bottom (Fig. 3.6b). This limits the box length to well below $2R$, the size of a corresponding cloud in the bare magnetic trap, as for larger boxes the harmonic trap would dominate. For the measurements presented in this thesis we therefore use box lengths between 38 and 60 μm.

A typical measured density profile of a quasi-condensate trapped in a box confinement is given in Fig. 3.6c. The parameters of the potential can be determined by recording density profiles for different atom numbers and fitting them with numerically obtained ground state profiles for a box trap with variable wall height and wall steepness. As the effective spot size of the transverse imaging system used to measure the profiles is larger than the expected wall extension it is not possible to precisely determine σ_{box}. We therefore assume $\sigma_{\text{box}} = 1.2\,\mu\text{m}$ which is the largest value still compatible with the observed profiles and agrees with the diffraction limit of the box imaging optics of 0.9 μm. With the magnetic confinement present, we only need wall heights that just surpass the chemical potential to fully confine the gas in the box. Tuning the laser power such that $U_0/h \simeq 1.3\,\text{kHz}$ suffices for typical parameters of the measurements presented in Chap. 5. The low intensities necessary to achieve these shallow potentials, together with the fact that for a blue detuned configuration the atoms mainly reside outside regions of high intensity, assures that scattering is low. For typical trap parameters we calculate a negligible scattering rate of $\Gamma_{sc} = 2\pi \cdot 10\,\mu\text{Hz}$ stemming from the overlap the of atomic density with the light field at the walls.

Previous implementations of optical dipole traps on atom chips focused only on optical lattices. They exploited the interference of retro-reflecting the dipole light from the chip surface [43] or used additional optical elements on the chip [44]. In our setup we would like to avoid such interference effects. However, the e^{-2} diameter of the dipole beam in the image plane is ~500 μm, which is much larger than the 100 μm distance of the atoms to the chip. Therefore, also light reflected from the chip passes the atoms and interference is inevitable, yet we do not see any unwanted effects from this.

Furthermore, sending a demagnified spot right on the chip surface should not be done without safety considerations. However, as we only apply very shallow potentials and the spot is spread out due to the small incidence angle, the intensity on the chip surface does not surpass a few mW/mm^2, which is far below any damage threshold. For a detailed study on chip heating effects due to reflected dipole trap light see Ref. [45].

A much more versatile dipole trap setup using a digital micormirror device (DMD) is discussed in Sect. 6.1. It is designed to enabling arbitrary longitudinal potentials, replacing the mask setup discussed here [46].

3.3 Probing

Various techniques to probe cold atom systems have been developed in the past decades [6]. Most of them exploit the interaction of atoms with light to gain information about their spatial distribution. Resonant or off-resonant light may be used for that purpose. For resonant light one can either detect the amount of photons absorbed by the atoms (absorption imaging), or directly detect the remitted fluorescence photons (fluorescence imaging). Using off-resonant light, the measurement of the local phase shift induced by the real part of the atom clouds index of refraction returns information about the atomic density distribution (phase contrast imaging). While the latter is interesting for probing high density samples and fluorescence imaging when single atom precision is desired, absorption imaging is the most straight forward and easily implementable technique. It is therefore the only technique used in the measurements described in this thesis.

As the atoms need to be released from the trap and the imaging process demands multiple scattering events heating up the cloud, absorption imaging is an inherently destructive technique. This means that the temporal evolution of a single realization of the system can not be tracked and each picture taken demands a repetition of the experimental cycle described in Sect. 3.2.1.

Three independent absorption imaging systems are implemented in the experiment. Before discussing their particularities in detail, the expansion of quasi-condensates after release from the trap will be described. Finally, a discussion of the most important measurement techniques used to obtain the results reported in Chaps. 4 and 5 is given.

3.3.1 Time-of-Flight Expansion

When the trap is turned off the atom cloud expands and gets accelerated away from the chip surface by gravity. Due to the tight transverse confinement the wave function expands radially much faster than in the weakly confined longitudinal direction. Therefore, let us first discuss the transverse expansion of a single gas. Assuming the atoms to be in the single particle ground state of the transverse harmonic potential, given in Eq. (2.31), expanding freely after the potential is switched off, we obtain [31]

$$\phi(x, y; t_{\text{tof}}) \propto \exp\left(-\frac{(x^2+y^2)}{2a_\perp^2 - i\frac{2\hbar}{m}t_{\text{tof}}} \right) = \exp\left(-\frac{(x^2+y^2)}{2\sigma_t^2} \right) \cdot \exp\left(i\frac{\hbar t_{\text{tof}}(x^2+y^2)}{2ma_\perp^2\sigma_t^2} \right), \tag{3.6}$$

with t_{tof} being the expansion time. The last step, separating the absolute value of the wave function from its phase, uses $\sigma_t^2 = a_\perp^2 + \left(\frac{\hbar t_{\text{tof}}}{ma_\perp}\right)^2$, giving the width of the wave function in expansion.

This calculation, however, neglects the role of transverse interactions which can have sizable effects on the clouds dynamics, as discussed in Sect. 2.2.2. Numerically solving the radial GPE for an expanding gas we can investigate these effects (see Appendix B). Figure 3.7a shows the transverse density profile during the initial expansion for typical parameters of our setup. After less than half a millisecond the central density drops to a fraction of its initial value. This is reflected in the fast drop of the interaction energy plotted in Fig. 3.7b. All energy initially stored in the interactions is transformed into kinetic energy in the first ~0.5 ms of the expansion. The repulsive interactions push the atoms apart and lead to a significant density dependence of the transverse width of the cloud (Fig. 3.7c). This is not captured by Eq. (3.6) which is independent of the density. Assuming the free expansion of a broadened Gaussian, as obtained in Sect. 2.2.2, would even lead to tighter density distributions at higher 1d densities due to the uncertainty relation between position and momentum (dashed green line in Fig. 3.7c). However, using the assumptions that the wave function is of Gaussian shape at all times we can obtain an expression for the width in expansion from simple energy arguments. Reformulating the time dependent term of σ_t^2 for

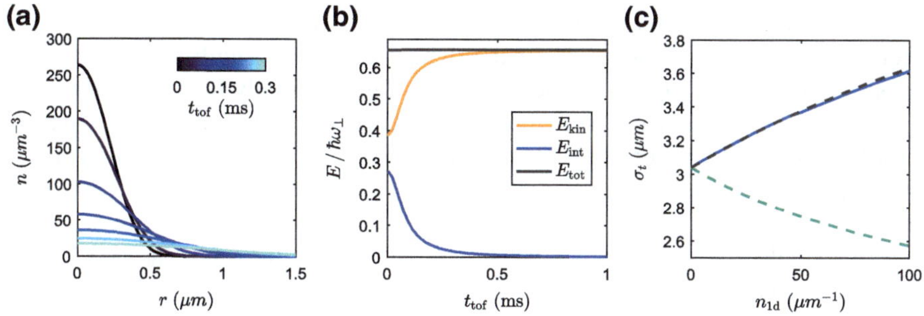

Fig. 3.7 Transverse expansion. **a** Evolution of the transverse radial density distribution after release from the trap for $n_{1\mathrm{d}} = 70\ \mu\mathrm{m}^{-1}$ and $\omega_\perp = 2\pi \cdot 2\,\mathrm{kHz}$, simulated by numerically solving the radial GPE (Appendix B). After 0.3 ms of expansion the central density dropped to ~7% of its initial value. **b** Dynamical transformation of interaction energy (blue) to kinetic energy (yellow) during the expansion. The total energy per particle is given in black. **c** Density dependence of the fitted wave function width after 1 ms of free expansion in the radial GPE (blue). The width obtained from the broadened Gaussian ansatz (Eq. 3.7) is shown by the dashed black line. If the interaction during the expansion would not be taken into account, the width would go down for higher densities, shown by the dashed green line

the free expansion to explicitly include the kinetic energy of the unbroadened single particle ground state $E_{\mathrm{kin}} = \frac{\hbar\omega_\perp}{2}$ and replaced it with $E_{\mathrm{kin}} + E_{\mathrm{int}} = \frac{\hbar^2}{2m\sigma^2} + \frac{\hbar^2 a_s n_{1\mathrm{d}}}{m\sigma^2}$ for the broadened wave function with $\sigma^2 = a_\perp^2\sqrt{1+2a_s n_{1\mathrm{d}}}$ we obtain[12]

$$\sigma_{t,n}^2 = \left[a_\perp^2 + \left(\frac{\hbar t_{\mathrm{tof}}}{m a_\perp}\right)^2\right]\sqrt{1+2a_s n_{1\mathrm{d}}}\,. \tag{3.7}$$

This means that the expanding width scales with density just as the width of the in situ wave function. Interestingly, it can be shown that for a variational approach using a broadened Gaussian to calculate the transverse expansion under interactions this expression provides a solution. It also agrees well with the results of the radial GPE simulations (Fig. 3.7c).[13] Note however that the broadening only enters in the absolute value of the wave function. The phase factor on the right side of Eq. (3.6) is unaffected.

Let us now turn to the longitudinal expansion dynamics. As the confinement along the longitudinal direction is typically much weaker the dynamical timescales are con-

[12]The expressions for E_{kin} and E_{int} in the broadened case are obtained by integrating over the transverse degrees of freedom for the kinetic and the interaction term of the Hamiltonian given in Eq. (2.30). Equation (3.7) was also found in [47].

[13]In Ref. [48] a similar density dependence of the width in expansion was measured experimentally, though the interpretation of the results has some shortcomings. No argument was given as to why it makes sense to compare the scaling of the measured density distribution width in expansion to the scaling of the in situ wave function width. Further, the expression for the width used was taken from [49], which reproduces the broadened chemical potential well but deviate significantly from the width found in [50] and the width obtained from numerical simulations of the radial GPE.

siderably longer.[14] Therefore, the initial period of hydrodynamic expansion on the order of $1/\omega_\perp$ is too short to affect the dynamics on the length scales accessible to our imaging. This means interaction effects can essentially be neglected and the expansion assumed to be ballistic. As a consequence, features in the recorded density distributions can easily be connected to properties of the trapped system. This is a major advantage of low dimensional cold gases working with single or few realizations.

For a single quasi-condensate expanding ballistically for times $t_{\text{tof}} \ll 1/\omega_z$, the average profile does not change significantly from the in situ density distribution.[15] However, the fluctuations present in the gas result in a near-field speckle pattern. This can be understood from calculating the mean field current density in the trapped gas for the phase-density representation of the wave function [51]

$$j(z) = \frac{\hbar}{m}\,\partial_z\theta(z)\left(1 + \frac{\delta n(z)}{n_{1\text{d}}}\right). \tag{3.8}$$

When the trap is turned off, this current distribution translates into longitudinal density fluctuations of the expanding cloud, referred to as density ripples. From Eq. (3.8) we see that these fluctuations are dominated by the gradient of the phase field. The in situ density fluctuations enter the current density only in a second order term and can be neglected.[16] Experimentally these patterns were first observed in [52] for elongated 3d condensates. Their strength, i.e. the amplitude of the ripple modulations for a given expansion time, depends on the magnitude of the in situ fluctuation and can therefore be used for thermomentry, as discussed later in Sect. 3.3.3.

In the case of a pair of gases released from a double well potential (see Sect. 3.1.3) the overlapping expansion leads to regular matter wave interference along the direction of the double well separation. Let us assume separable wave functions $\Psi_i(\mathbf{r}) = \psi_i(z)\phi_i(x, y)$ with $i = 1, 2$ indexing the two wells separated by the distance d along the x-axis. All atoms occupy the transverse single particle ground state (Eq. 2.31) of their respective well. Calculating the density distribution for a free expansion in transverse directions, employing the phase-density representation of $\psi_i(z)$ (Eq. 2.5), we obtain

$$\begin{aligned} n(\mathbf{r}) &= |\Psi_1(\mathbf{r}) + \Psi_2(\mathbf{r})|^2 \\ &= |\psi_1(z)|^2|\phi_1(x,y)|^2 + |\psi_2(z)|^2|\phi_2(x,y)|^2 + 2\text{Re}\left[\psi_1^*(z)\psi_2(z)\phi_1^*(x,y)\phi_2(x,y)\right] \\ &\propto e^{\frac{x^2+y^2}{\sigma_t^2}}\left[|\psi_1(z)|^2 + |\psi_2(z)|^2 + 2\text{Re}\left(\psi_1^*(z)\psi_2(z)e^{-i\frac{md}{\hbar t_{\text{tof}}}x}\right)\right] \\ &\propto e^{\frac{x^2+y^2}{\sigma_t^2}}\left[n_1(z) + n_2(z) + \sqrt{n_1(z)n_2(z)}\cos\left(\theta_1(z) - \theta_2(z) + \frac{md}{\hbar t_{\text{tof}}}x\right)\right]. \end{aligned} \tag{3.9}$$

---

[14]For the harmonic longitudinal confinement typically $\omega_\perp / \omega_z \sim 200$.

[15]For the box trap it would be $t_{\text{tof}} \ll L/c\pi$.

[16]Note that the influence of the in situ density fluctuations on the final ripple pattern mainly stems from their modulation of the initial pre-expansion density profile.

Here, we further assumed the expansion time to be long enough such that $\sigma_t \gg a_\perp$ and therefore $\sigma_t \approx \frac{\hbar t_{\text{tof}}}{m a_\perp}$. The time dependencies of ψ_i, ϕ_i and n_i are omitted for clarity. From the cosine factor in Eq. (3.9) we see that in expansion interference fringes along the x-direction emerge whose phase is modulated by the relative phase between the condensates $\varphi(z) = \theta_1(z) - \theta_2(z)$. As discussed later in Sect. 3.3.4, this allows for a direct measurement of the in situ relative phase field which is the central observable for the recurrence dynamics discussed in Chap. 5.

The spacing of the fringes along x is given by

$$\lambda_{\text{fs}} = \frac{2\pi\hbar t_{\text{tof}}}{md} .$$

As the phase factor in Eq. (3.6) is not affected by the broadening of the transverse wave function the fringe spacing is also not influenced by transverse interaction effects. Only the size of the entire pattern will be affected by $\sigma_{t,n}$, entering the Gaussian pre-factor in Eq. (3.9). This translates into a slight density dependence of the number of fringes visible.[17]

Note that longitudinally the same density ripple patterns appear as discussed earlier for a single gas. However, here we observe an incoherent sum of two of these patterns originating from the two clouds.

3.3.2 *Absorption Imaging*

After expansion the atomic density distribution is detected via absorption imaging. The cloud is illuminated by laser light resonant to the $F=2 \rightarrow F'=3$ transition of the ^{87}Rb D_2 line while a CCD records the amount of light the atoms scatter from the beam. Characterizing the gain and quantum efficiency of our cameras allows for absolute intensity measurements lifting the restriction to intensities far below saturation [53]. However, we still only use intensities between 15 and 35% of the saturation intensity. For a detailed discussion on the role of the imaging quantization axis, the light polarization and the effects of stray fields the reader is refereed to Ref. [54].

As absorption imaging projects the 3d density distribution of the expanded cloud onto a 2d image it intrinsically integrates the profile along the imaging axis. Therefore, different properties of the gas can be explored via different imaging axes, leading to the implementation of three independent absorption imaging systems in our setup. They are oriented roughly perpendicular to each other along the major axis of the trap, as sketched in Fig. 3.8. In the following, each imaging system will be discussed briefly, focusing on the aspects relevant to the results presented in this thesis. For more detailed information on the setups the reader is referred to [31, 34, 36].

---

[17] Calculating the number of fringes within the full $1/e^2$-width we obtain $\frac{2\sqrt{2}\sigma_{t,n}}{\lambda_{\text{fs}}} = \frac{\sqrt{2}}{\pi}\frac{d}{a_\perp}\sqrt[4]{1+2a_s n_{1\text{d}}}$.

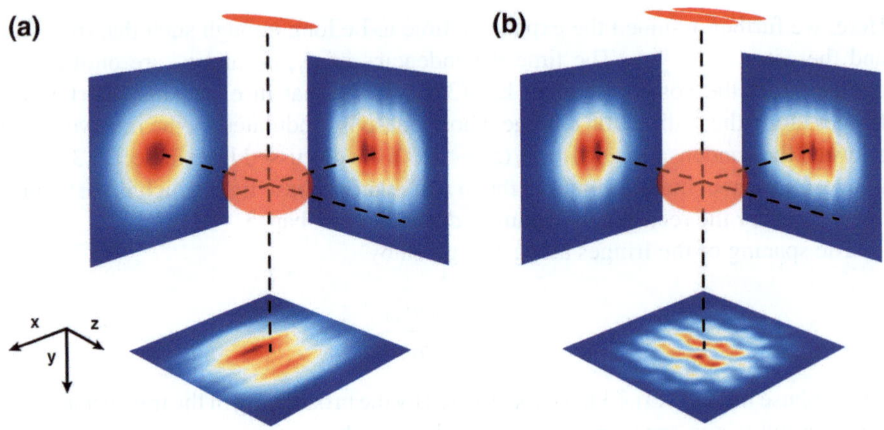

Fig. 3.8 **Illustration of the accessible imaging axes.** **a** Schematic depiction of the expansion of a single 1d quasi-condensate (red) and the orientation of the imaging axes (dashed lines) with respect to the cloud geometry. The panels perpendicular to the imaging axes show density distributions integrated along the respective axes. The left panel shows the perspective accessible to the longitudinal imaging (integrated along the z-axis), the right panel the one accessible to the transverse imaging (integrated along the x-axis) and the lower panel the one accessible to the vertical imaging (integrated along the y-axis). The plotted example distributions are obtained from simulations. **b** The same for two adjacent 1d quasi-condensate released from a double well potential, leading to interference patterns visible in the longitudinal and vertical imaging. Figure adapted from Refs. [31, 33]

In the following discussion note that the values for the imaging resolution used throughout this thesis are always to be understood as effective. When imaged, the atomic cloud is generally larger then the depth of field of the imaging system, and additionally gets pushed through the focal region by the imaging beam. This leads to a smearing of the point spread function and thereby to an effectively larger spot size. The size of this effective spot depends on the density via the cloud width, the imaging intensity and the exposure time. For a detailed discussion of these effects see Ref. [54]. Throughout the thesis the effective spot size in intensity is referred to by the width of its Gaussian approximation σ_{psf}.

Transverse Imaging System

This imaging system is aligned transverse to the z-axis of the chip trap i.e. the 1d axis of the cloud. It is oriented at a 2° angle to the chip surface such that the imaging light is reflected from the surface to allow for the measurement of clouds after very short expansion times (Fig. 3.9a). For longer expansion times, $t_{\text{tof}} > 10$ ms, the clouds are far enough from the chip that the imaging light passes below the chip without reflection. The imaged 3d density distribution is integrated along an axis almost aligned with the x-direction. Therefore, the observed profiles provide access to density ripple patterns discussed in Sect. 3.3.1, forming along the 1d axis during expansion. The right panel in Fig. 3.8a shows an example of such a pattern obtained from simulations while later Fig. 3.10a shows examples of measured density patterns.

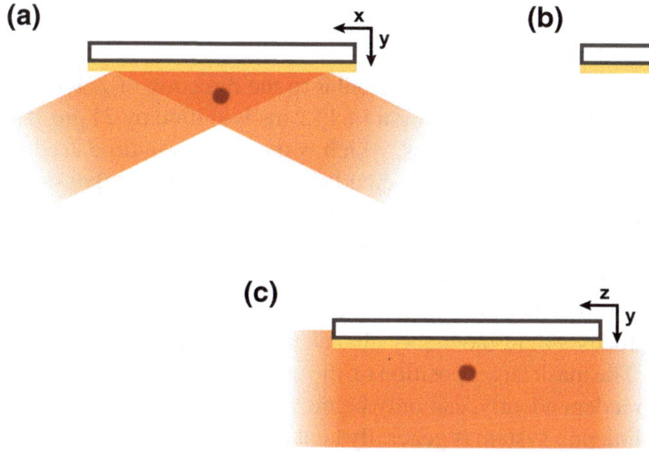

Fig. 3.9 Schematic of the imaging light paths. **a** Light path of the transverse imaging system for short expansion times, reflecting from the chip surface. The angle between the imaging axis and the chip is exaggerated. For longer expansion times the light path passes below the chip as in (**c**). **b** Beam alignment for the vertical imaging system. **c** Beam alignment for the longitudinal imaging system

In the case of two gases released from a double well trap (Fig. 3.8b) we observe the same since the wells are separated along x such that all interference effects are integrated out by the imaging process.

The core of the imaging system is a home built four lens objective with a numerical aperture (NA) of 0.26. Together with a two lens telephoto combination on the image side, this system magnifies the image of the observed density distributions by a factor of 12.4 onto a CCD camera.[18] The pixel size in object space is 1.05 µm and the field of view spans 1.07 mm × 0.72 mm, allowing us to observe the falling clouds for up to 12 ms. An additional adjustable aperture set to a diameter of ~36 mm in the parallel part of the imaging system limits the effective NA to 0.20 in order to minimize aberrations. This leads to a diffraction limited intensity spot of width $\sigma_{\text{psf}} = 0.9\,\mu\text{m}$, when approximated by a Gaussian. Due to the small depth of field resulting from the large NA, the resolution is drastically reduced by the spatial extension of the cloud and the push through the focal plane. For typical cloud sizes and imaging parameters we measure an effective spot size of $\sigma_{\text{psf}} \simeq 2.6\,\mu\text{m}$ from the minimum position of the density ripple correlations (see Sect. 3.3.3). This signal can also be used to focus the imaging system. While coarse focusing is achieved by intentionally detuning the imaging light by a few MHz and minimizing the diffraction effects caused by a dense cloud of atoms [55], the density ripple correlations provide a more sensitive probe giving discernable signals down to focus differences of around 10 µm for clouds that

[18] Andor DV435-BV-958.

extend $\sim 100\,\mu m$ along the imaging axis.[19] Further, also the power spectrum of the density ripples can be used for focusing [56].

As the imaging axis is not perfectly perpendicular to the direction of gravity the focal point found for the standard time-of-flight of 11.2 ms is shifted by 21 μm with respect to the cloud position after $t_{tof} = 2$ ms, which is the expansion time used for (quasi) in situ imaging. Also, as the transverse cloud size after this short expansion is much smaller and the push through the focal plane is not linear the optimal focal point to minimize the effective spot size will be different. Using an adjusted focus for short expansion times is however not feasible for the in situ measurements presented in this thesis (see Figs. 3.6c and 5.6) as they also involve the dipole box trap (see Sect. 3.2.3). Changing the imaging focus would inevitably also change the dipole trap focus as they share an objective. The mask target position on the other hand, which can be used to focus the dipole trap independently, can only be moved manually. Therefore, the focus of the transverse imaging system is generally kept aligned to the $t_{tof} = 11.2$ ms cloud position. For $t_{tof} = 2$ ms the slight defocusing leads to an effective spot size similar to the one obtained for $t_{tof} = 11.2$ ms even though the smaller cloud width would in principle result in a better resolution.[20] Another difficulty when working with short expansion times is that the reflection of the imaging light from the chip surface leads to a standing wave pattern at the point of the atoms as well as an additional virtual image that needs to be accounted for [28]. Hence, all density profiles presented in this thesis that were measured with short expansion times were rescaled to fit the expected total number of particles independently measured at the standard time-of-flight.[21]

The dipole box trap discussed in Sect. 3.2.3 is imaged onto the atoms through the objective of the transverse imaging system. Therefore, stray light of the dipole trap would inevitably scatter into the imaging system. As the camera does not have a shutter and the exposure is controlled by an AOM switching the imaging light, this dipole light leads to unwanted counts on the CCD. The camera applies a cleaning shift when no picture is taken, continuously clearing the CCD of stray light counts, however, this shift is quite slow. It takes 24 ms to completely clear the uncovered part of the CCD. This means already minuscule intensities around $1\,nW/mm^2$ sum up to counts comparable to the ones recorded in a typical imaging situation. To avoid the unwanted exposure of the CCD we use two narrow bandpass filters[22] centered at 780 nm. They are placed right in front of the CCD, spaced ~8 cm apart to minimize reflections between them.[23] With these filters no effect of the dipole light on the

---

[19]The strategy here is to maximize the g_2 contrast C_{g_2} or to find the smallest minimum position δz_{min} for a given expansion time. See Sect. 3.3.3 for a definition of these quantities.

[20]This was concluded only from simulations as a sensitive measurement of the effective spot size, as through the density ripples correlations in the case of long expansion times, is not possible.

[21]Possible non-linearities due to different saturation levels are neglected.

[22]Semrock SEM-LL01-780-25.

[23]Even though the filters are mounted parallel to each other, most of the dipole light probably hits the filters at a small angle. For a large separation this causes the light to undergo only a small number of reflections before exiting the inter-filter cavity and being dumped, leading to a higher total optical density.

transverse imaging process can be observed allowing also for very short times-of-flight where the dipole trap is turned off only a few milliseconds before the picture is taken.

The tilt of the transverse imaging axis with respect to the chip surface is realized by a tilted mounting of the entire bread board holding the optics and camera. Due to this the bread board is not as tightly connected to the surrounding experimental structures, making it more prone to vibrations. To minimize the diffraction structures entering the absorption images through these vibrations we employ a fringe removal post-processing technique that calculates optimal background images from all background images taken [57]. Especially for small atom numbers, where the signal to noise ratio is low, this technique greatly improves the image quality. It is used primarily in the measurements presented in Chap. 4.

Vertical Imaging System

For imaging along an axis (nearly) aligned to gravity the chip poses a major obstacle. As for the transverse imaging, the imaging light can be reflected from the chip surface, however, the additional unfocused virtual image obscures the real image. To avoid this, while still maintaining a near-vertical orientation of the imaging axis (along the y-axis), we focus the imaging light close to the position of the atom cloud, slightly displaced in the xz-plane. Thereby, we bypass the atoms before the light is reflected and restrict atom-light interactions to the beam bounced off the chip surface (Fig. 3.9b). For single quasi-condensates, images taken along this axis show the same ripple patterns as the transverse imaging system (Fig. 3.8a) but for two gases released from a double well trap the vertical imaging provides access to the full interference pattern (Fig. 3.8b).

The optical setup of this imaging system is composed of a single doublet objective lens and a two lens telephoto combination casting the image onto a CCD camera.[24] The imaging light enters the vacuum chamber through the same objective lens and is fed into the setup through a 1.5 inch polarizing beam splitter cube. With a magnification of 8.22 we obtain a pixel size in object space of 1.95 μm and a field of view of 1 mm × 1 mm. Since the atoms fall towards the objective during expansion the focus needs to be adjusted to the desired time-of-flight. However, this is complicated by an astigmatism of the optical setup due to a slight tilt of the imaging light path with respect to the optical axis. As mentioned before, the tilt is introduced to focus the imaging light beside the atoms, thereby avoiding the virtual image. Using different features along the x and z-axis to find the two foci independently we can quantify this aberration and account for it. Along the x-axis, where we can observe the interference fringes, we use the detuning technique discussed for transverse imaging system. This works particularly well on clouds showing interference patterns. When defocused, the atomic density distribution acts as a grating such that the observed interference contrast depends on the laser detuning [31]. In focus, however, the contrast is independent of the detuning. Along the z-axis, the density ripples can be used, as for the transverse imaging system. The foci obtained from these two methods differ by

[24] Andor iXon DV887DCS-BV. An EMCCD camera used without the electron multiplier mode.

$\sim$ 120 μm. From the NA of 0.08 we would expect a diffraction limited spot size of 2.2 μm, without the astigmatism. In practice, the effective spot size for the pushed and extended cloud will depend on the chosen focus and the feature of interest.

The vertical imaging CCD needs to be shielded from stray dipole trap light as well. As the dipole beam is reflected on the chip surface, some light is scattered from the etched trenches between the wires and impurities on the chip surface. To prevent this light from reaching the CCD a shutter is installed right in front of the camera. With the region of interest on the CCD only about 2 mm wide we are able to use a small aperture shutter[25] with low vibrations, not affecting the imaging process.

Longitudinal Imaging System

The axis of the third imaging system is aligned with the weakly confining axis of the chip trap. Hence, the imaging process integrates the expanded 3d density distributions along the z-axis and returns the distribution of particles in the xy-plane. This means that primarily information about the transverse degrees of freedom are accessible with this setup. In the left panel of Fig. 3.8a we see the transverse Gaussian density distribution of an expanding quasi-condensate, as given in Eq. (3.6). For gases released from a double well trap interference can be observed. The integration along the 1d axis smears out the fringes for quasi-condensates with strong relative phase fluctuations and returns high contrast only for phase coherent samples (Fig. 3.8b, left panel).

The imaging system consists of two doublet lenses and a CCD camera.[26] Its magnification is 5.3, leading to a pixel size in object space of 2.45 μm. The field of view spans 2.51 mm × 1.67 mm allowing us to observe a falling cloud of atoms from its release up to $\sim$18 ms of expansion time. As the imaging light passes close to the chip (Fig. 3.9c) diffraction effects from the chip mounting distort the images for short expansion times $t_{\text{tof}} < 4$ ms. Focusing can be achieved by applying the detuning technique discussed for the other two imaging systems, ideally on a cloud exhibiting interference fringes.

When used in combination with the dipole trap the longitudinal imaging is limited to $t_{\text{tof}} < 11$ ms due to the light scattered from the chip structures. For shorter times-of-flights the cleaning shift of the camera is fast enough to clear the region of interest. No mechanism to avoid the dipole light reaching the CCD is installed.

3.3.3 Density Ripple Thermometry

As discussed in Sect. 3.3.1, the quasi-condensate fluctuations translate into 1d density ripple patterns when the clouds expand after their release from the trap. Figure 3.10a shows the formation of such patterns, recorded by the transverse imaging system. Since the magnitude of the ripple modulations directly relates to the strength of the

---

[25] Uniblitz LS6. This shutter takes 0.7 ms to fully open its 6 mm wide aperture.

[26] Andor DV435-BV-958.

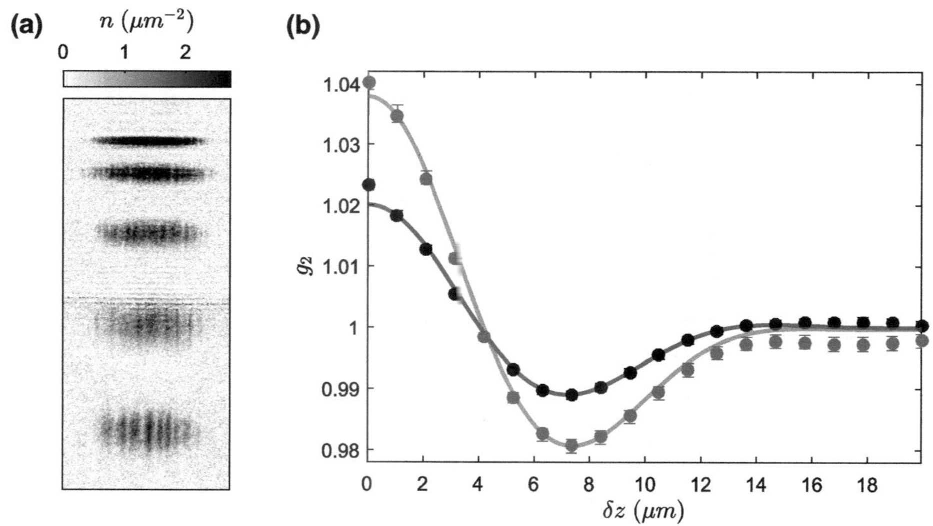

Fig. 3.10 Density correlation measurement. **a** Absorption images of single quasi-condensates for different expansion times combined in a single image ($t_{\text{tof}} = 2, 4.4, 6.8, 9.2,$ and$11.6\,\text{ms}$). The clouds represent independent realizations showing uncorrelated ripple patterns. The horizontal fringes visible in the center of the image are due to diffraction on the chip edge. **b** Two examples of measured g_2 functions, with the error bars giving the 68% confidence intervals obtained from a bootstrap (Appendix C). The measurements were performed on pairs of strongly coupled gases confined in a box trap, exhibiting an inferred thermal coherence length of $\lambda_T = 24.5\,\mu\text{m}$ (black) and $\lambda_T = 9.5\,\mu\text{m}$ (gray). The solid lines show the corresponding correlations obtained from simulated data. A small global offset of 1–2% of C_{g_2} was added to the measured data to counteract normalization errors and better show the agreement with the simulated data

in situ fluctuations (see Eq. 3.8), they can serve as a means of thermometry. This technique was first developed in Ref. [58] and experimentally applied in Ref. [59]. It provides a thermometry signal in situations where the standard method of fitting the expansion of the thermal background atoms fails due to a negligible thermal fractions or insufficient expansion time. Also, it directly measures the condensate temperature such that no assumption of equilibrium between the thermal and the condensed fraction is necessary.

To quantify the observed density modulations we calculate the normalized two-point density correlation function

$$g_2(\delta z) = \frac{\int \mathrm{d}z\, \langle n(z+\delta z) n(z)\rangle}{\int \mathrm{d}z\, \langle n(z+\delta z)\rangle \langle n(z)\rangle}, \tag{3.10}$$

where the expectation values $\langle \dots \rangle$ are taken over many experimental realizations of recorded profiles $n(z)$. In Fig. 3.10b, examples for these g_2 functions are shown, calculated from measured density profiles. A sample size of at least 100 realizations generally suffices to obtain a reliable signal. These measured g_2 functions are then compared to equilibrium predictions in order to estimate the condensate temperature.

For a homogeneous weakly interacting Bose gas Ref. [58] provides an analytic expression of the ripple power spectrum that can be directly compared to experimentally observed spectra. However, calculating the emerging ripple patters from numerically sampled in situ fluctuation (see Sect. 2.3.1) allows to take the inhomogeneous density profile and the transverse broadening into account. To this end, we generate on the order of 10^5 thermal wave functions within the quadratic Luttinger liquid approximation and propagate them freely for the expansion time t_{tof} (Fig. 3.11a). The expanded density profiles are then convolved with an effective Gaussian point spread function to take the experimental imaging resolution into account. From the obtained set of smeared out profiles the g_2 function is calculated in the same way as for the experimental samples (Eq. 3.10). Figure 3.11b, c and d show examples of the obtained correlations for different imaging resolutions, thermal coherence lengths and expansion times, respectively.

From measured g_2 functions the thermal coherence length is inferred by comparing them to these simulated correlations. This can be achieved by a direct fit, with λ_T kept as a free parameter. However, looking closely at the behavior of the g_2 function under changes of λ_T (Fig. 3.11c) we observe that the relevant features are the height of the peak[27] at $\delta z = 0$ and the depth of the minimum at δz_{min}. It turns out that the difference of these values $C_{g_2} = g_2(0) - g_2(\delta z_{\mathrm{min}})$, referred to as the g_2 contrast, is indirectly proportional to the thermal coherence length, $\lambda_T = a^{(0)} + a^{(1)} \cdot C_{g_2}^{-1}$ (Fig. 3.11e).[28] Extracting the proportionality factors $a^{(1,2)}$ from the simulated correlations we can therefore estimate λ_T from the experimentally observed contrasts. First described in Ref. [60], this technique is faster and more robust than the full fit. It is insensitive to

[27] The peak height was the observable used for thermometry in [59].

[28] Numerically we observe that this relation breaks down when the size of the density modulations emerging in expansion get comparable to the average density.

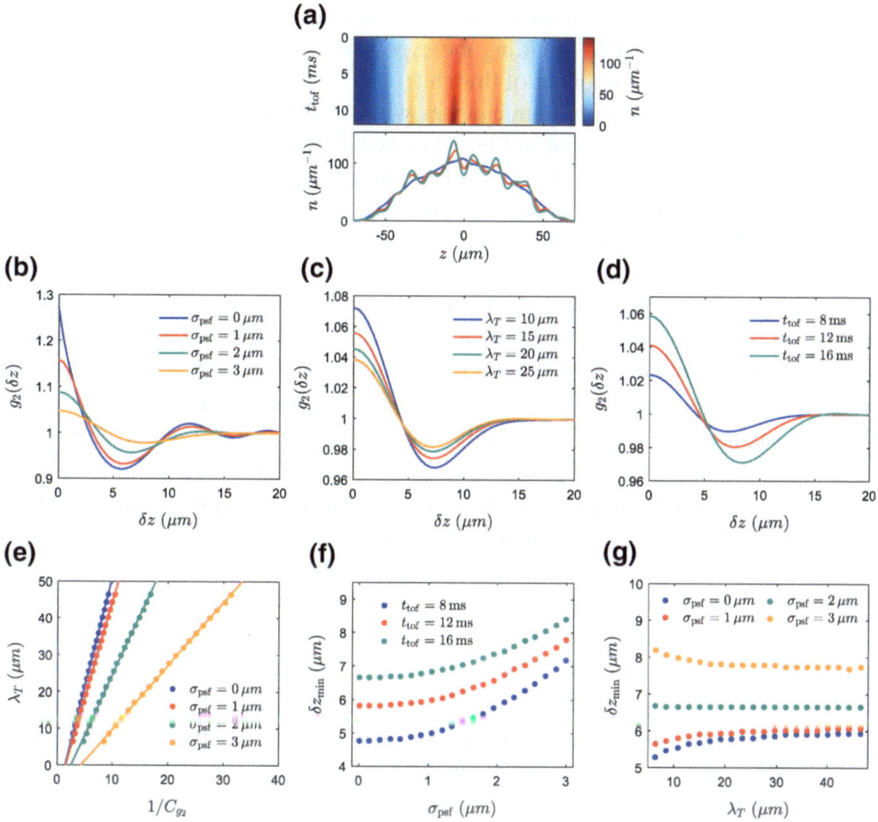

Fig. 3.11 Simulated density ripple correlations. a Ballistic expansion dynamics calculated for a single wave function realization sampled from a thermal state of the Luttinger liquid Hamiltonian at $T = 40$ nK (Sect. 2.3.1). While the upper panel gives the full expansion dynamics, the lower panel shows cuts at $t_{tof} = 0, 6, 12$ ms (blue, red, green). In both, the typical imaging resolution is taken into account. **b** Effect of the limited imaging resolution on the g_2 functions. For larger effective point spreads σ_{psf} the g_2 contrast decreases and δz_{min} shifts to larger distances. **c** Correlations for different thermal coherence lengths λ_T. A larger coherence length leads to a smaller g_2 contrast. **d** Effects of the expansion time t_{tof} on the g_2 functions. **e** Linar relation between λ_T and the inverse g_2 contrast $C_{g_2}^{-1}$. **f** Shift of the g_2 minimum position δz_{min} with σ_{psf} for different expansion times t_{tof}. **g** Minimum position δz_{min} over λ_T for different σ_{psf}

normalization shifts and other distortions and allows for a fast resampling, necessary to calculate confidence intervals from a bootstrap (Appendix C).

Experimentally, the shot noise of the imaging light enters in $g_2(0)$. In Fig. 3.10b this can be observed when comparing the data points at $\delta z = 0$ with the corresponding simulated data. We therefore use the second point of the measured g_2 functions to extract C_{g_2} and take this into account in the comparison to the simulated data. The minimum value $g_2(\delta z_{min})$ is extracted from a third-order polynomial fit around the lowest points to surpass the discreteness of the measured data. Further, experimental shot-to-shot fluctuations of the total atom number lead to large normalization shifts, as expected from Eq. (3.10). Dividing the experimental g_2 functions by $\langle N^2\rangle/\langle N\rangle^2$ removes the majority component of this shift.

The position of the g_2 minimum changes with the effective imaging point spread and the expansion time, as seen in Fig. 3.11b and d. Both longer t_{tof} and larger σ_{psf} shift δz_{min} to larger distances (Fig. 3.11f). However, δz_{min} is almost insensitive against changes in λ_T which makes it an ideal probe for the effective imaging resolution (Fig. 3.11g). Also, focusing can be achieved by minimizing δz_{min}, or maximizing C_{g_2}, as mentioned in Sect. 3.3.2.

For pairs of gases trapped in a double well potential the same ripple patterns form in expansion. If uncoupled, the observed pattern results from the incoherent sum of two independent patterns. In that case, the resulting $g_2^{(J=0)}$ relates to that of a single gas at the same temperature as $g_2^{(J=0)} = (1 + g_2)/2$, corresponding to a reduction of C_{g_2} by a factor of two. For strongly coupled gases, the phase fluctuations are locked such that the resulting correlations $g_2^{(J=\infty)}$ are equal to that of just one of the two. However, as the common phase fluctuations in two coupled gases are of the same magnitude as the phase fluctuations in a single gas with the same coherence length, $2\langle|\varphi_{c,k}|^2\rangle \simeq \langle|\theta_k|^2\rangle$, and the expectation value of the squared phase mode amplitudes is approximately proportional to C_{g_2}, the correlations in the coupled case are as large as in the uncoupled case, $g_2^{(J=\infty)} \approx g_2^{(J=0)}$. Therefore, in both cases the direct proportionality factor between λ_T and $C_{g_2}^{-1}$ is simply given half of that of a single gas, $a_J^{(1)} = a^{(1)}/2$.

Note that for the simulated data presented here we used the discretized Luttinger liquid, taking both density and phase fluctuations into account (see Sect. 2.3.1). In most experimentally relevant cases, however, the phase fluctuations dominate the observed patterns and the density fluctuations can be neglected. A simple stochastic process [61] generating phase fluctuation patterns for inhomogeneous background densities can therefore be used as well to simulate thermal g_2 functions.

3.3.4 Relative Phase Measurement

From the interference term in the density distribution of two expanding quasi-condensates (Eq. 3.9) we see that the phase of the interference pattern is modulated along the z-axis by the relative phase field between the two gases, $\varphi(z) = \theta_1(z) - \theta_2(z)$. Fluctuations in the relative phase therefore directly translate into fluctuations of the interference fringe positions. The vertical imaging system allows us to image this modulation and extract $\varphi(z)$ (Fig. 3.12). For this we individually fit each pixel row along z with a cosine modulated Gaussian, as shown in Fig. 3.12b for three examples. In these fits we extract the contrast (i.e. the amplitude of the cosine modulation) and the phase, while keeping the fringe spacing and the Gaussian center position fixed. The fringe spacing is extracted separately and averaged over the whole ensemble of images taken with the same configuration. The central position is extracted from a separate Gaussian fit of the image binned along the z-axis.

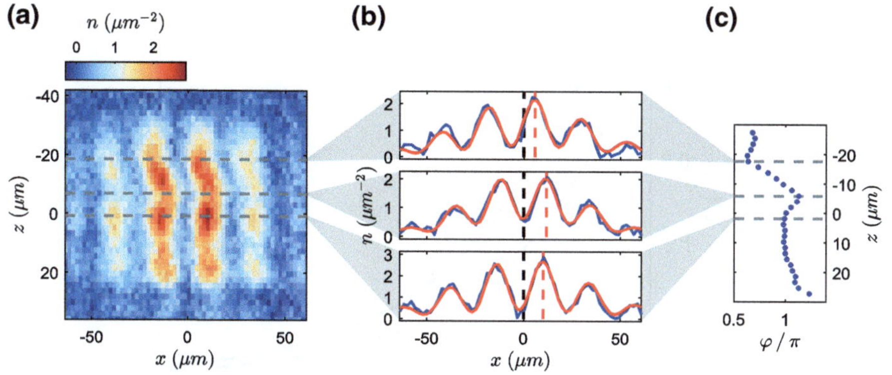

Fig. 3.12 Extraction of the relative phase field. a Absorption image of the matter-wave interference pattern generated by two 1d superfluids in expansion. The 1d axis of the gases is aligned along z while the double well separation runs along x. **b** Selected pixel rows of the interference pattern (blue) shown with the corresponding fits (red). The dashed red line indicates the position of the fitted most central fringe. Would the local relative phase be zero this line would coincide with the dashed black line at $x = 0$. **c** Relative phase field $\varphi(z)$ extracted from the pattern in (**a**) by individually fitting each pixel row. Unphysical jumps $> \pi$ are removed

The ability to extract the entire relative phase field enables us to analyze its spatial correlations and their dynamics [62–64]. In this thesis, we are mainly interested in the phase correlation function, as defined in Eq. (2.25) of Sect. 2.1.3. The dynamics of this correlation function after decoupling two gases trapped in a longitudinal box confinement is the topic of Chap. 5. However, when comparing measured phase profiles or correlation functions to theoretical predictions, like the thermal correlations given in Eqs. (2.27) and (2.28), we need to consider the finite imaging resolution. As the images of the interference patterns are subject to a certain resolution limit also the extracted phase fields are smeared over some effective point spread length σ_{psf} (see Ref. [54]). Assuming this spread to be Gaussian we can take it into account when calculating $C(\bar{z})$ for states with Gaussian fluctuations, as in Eq. (2.26), and obtain

$$C(\bar{z}, \sigma_{\text{psf}}) = \exp\left[-\int_0^\infty \frac{\mathrm{d}k}{\pi} \langle|\varphi_k|^2\rangle\, e^{-k^2\sigma_{\text{psf}}^2} \left(1 - \cos(k\bar{z})\right)\right]. \tag{3.11}$$

Inserting the equilibrium phase mode amplitudes of two uncoupled gases this gives

$$C(\bar{z}, \sigma_{\text{psf}}) = \exp\left[-\frac{2\bar{z}}{\lambda_T}\left(\text{erf}\left(\frac{\bar{z}}{2\sigma_{\text{psf}}}\right) + \frac{2\sigma_{\text{psf}}}{\bar{z}\sqrt{\pi}}\left(e^{-\left(\frac{\bar{z}}{2\sigma_{\text{psf}}}\right)^2} - 1\right)\right)\right], \tag{3.12}$$

which, for large distances $z \gg \sigma$, approaches the ideal unsmeared form given in Eq. (2.27) (Fig. 3.13). Figure 3.14a shows examples for measured phase correlation functions of uncoupled systems and compares them to thermal fits employing Eq. (3.12).

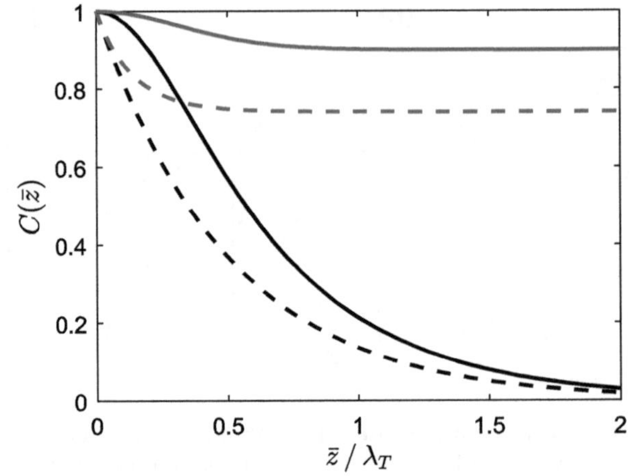

Fig. 3.13 Finite resolution effects on the phase correlation function. Comparison of equilibrium phase correlation functions (dashed lines) with the ones obtained from phase profiles smeared by a finite resolution imaging process (solid lines). The plot shows the case of two uncoupled gases (black) and two strongly coupled gases with $l_J = 0.15 \cdot \lambda_T$ (gray). The effective imaging resolution is given by $\sigma_{\text{psf}} = 0.2 \cdot \lambda_T$

Such fits provides a method of thermometry for the relative degrees of freedom. For coupled gases in thermal equilibrium (Fig. 3.13) or non-equilibrium correlations, Eq. (3.11) can be evaluated numerically.

The effective resolution of the extracted phase profiles is strongly influenced by the astigmatism of the vertical imaging system discussed in Sect. 3.3.2. The question arises for which focus the phase fluctuations are optimally resolved. While the relative phase is a quantity changing along z, a fit of the fringes along x is used to extract it. To find the optimal focus we measure the phase correlations function of two uncoupled quasi-condensates in thermal equilibrium for different times-of-flight (Fig. 3.14b). The different expansion times translate into different cloud positions with respect to the foci along the x and z-axis, as the cloud falls towards the objective. Fitting the phase correlations with Eq. (3.12) keeping σ_{psf} a free parameter we find that the optimal focus for the extraction of phase profiles lies in between the one obtained from focusing the fringes along x and the one found focusing the density ripples along z (Fig. 3.12c). All measurements presented in this thesis therefore use this effective focus. Comparing phase correlations with predictions for thermal equilibrium states, as in Fig. 3.14a, we estimate an effective spot size of $\sigma_{\text{psf}} \simeq 3.0\,\mu\text{m}$.

For higher temperatures, the finite imaging resolution renders the phase extraction partially unreliable. With λ_T no longer much larger than σ_{psf}, large phase fluctuations on length scales of the resolution limit become possible. If the phase winds close to 2π or more on such short scales the contrast on the corresponding pixel rows will go to zero and the fit will locally fail. For the extracted phase profiles this introduces an effective cut-off for large short range fluctuations. Simulations of the imaging process show that for thermal coherence lengths below $\lambda_T \sim 10\,\mu\text{m}$ the thermometry through

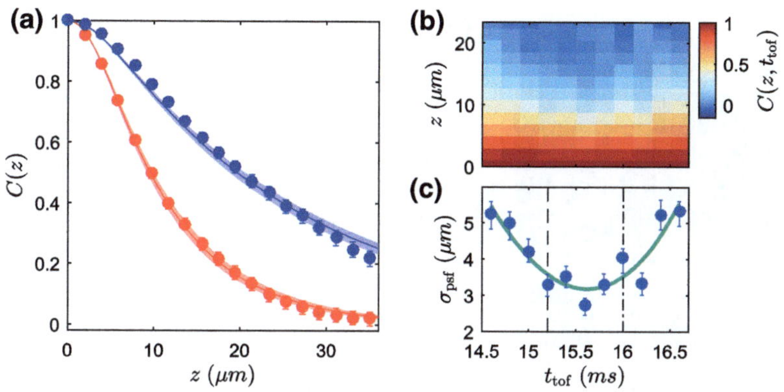

Fig. 3.14 Measured phase correlation functions. a Two examples for measured equilibrium phase correlation functions of uncoupled gases. The error bars give the 68% confidence interval obtained from a bootstrap (Appendix C). Fits with Eq. (3.12) are shown as solid lines, returning $\lambda_T = 47.3\,\mu\text{m}$ ($44.7\,\mu\text{m}$, $49.8\,\mu\text{m}$) and $\lambda_T = 18.0\,\mu\text{m}$ ($17.0\,\mu\text{m}$, $19.0\,\mu\text{m}$) respectively for the blue and red data points. The confidence intervals of the fit indicated by the shaded areas are obtained as well from a bootstrap. **b** Phase correlation function for different times-of-flight. **c** Effective spot size extracted from the data shown in (**b**) assuming a constant temperature. Again a bootstrap is used to extract the confidence intervals. The solid green line shows a quadratic fit indicating that the effective focus for the phase extraction lies between the density ripple focus (dashed line) and the fringe focus (dash-dotted line). The temperature in this measurement was too high for faithful thermometry therefore the absolute values of the extracted $\sigma_{\rm psf}$ can not be trusted

fitting the phase correlation function starts to fail. With typical densities of $n_{1\rm d} = 70\ \mu\text{m}^{-1}$ this corresponds to a temperatures of $\sim$70 nK. However, temperatures above this limit can still be faithfully inferred from the contrast distributions, as discussed in the following section. For phase correlations between points separated by large distances, like the recurrence signal discussed in Chap. 5, overlooking short range jumps of 2π does not matter.

3.3.5 *Contrast Distribution Thermometry*

Another technique to investigate the spatial correlations of the relative phase between two quasi-condensates is to extract the distribution of interference contrasts. Integrating the interference patterns obtained from the vertical imaging system along z and fitting them with a cosine modulated Gaussian, as shown for the individual pixel rows in Fig. 3.12b, returns the contrast C as the amplitude of the modulation. A flat relative phase field resulting in straight fringes will lead to a high contrast while strong fluctuations in the phase wash out the integrated interference pattern and lead to a vanishing contrast. Hence, the distribution of observed contrasts contains information about the degree of phase correlation within the integrated region. Performing the integration over different length scales adds the spatial component

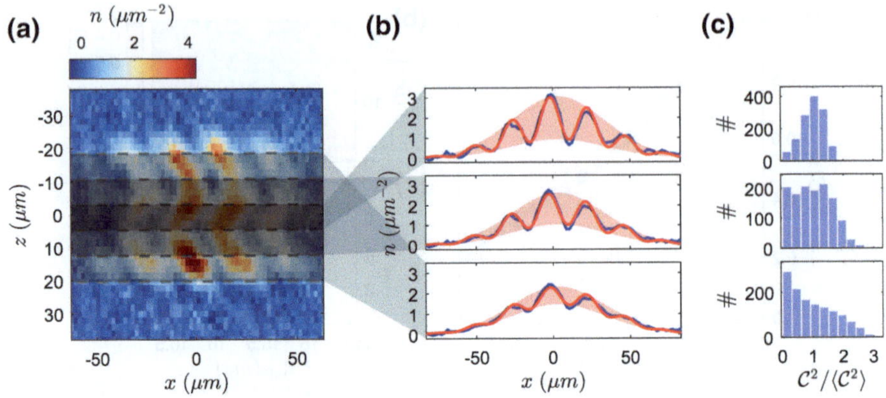

Fig. 3.15 Extraction of interference contrasts. a Matter-wave interference pattern with three integration regions indicated by the shaded areas enclosed in the dashed gray lines. Integration is performed along the z-axis over the length of $s = 9.7$, 25.3, and 40.9 µm, respectively. **b** Integrated interference patterns (blue) together with the corresponding fits (red). The width of the red shaded area indicates the fitted interference contrast $\mathcal{C}$. The pattern in (**a**) is summed up and divided by s to allow for comparisons between the panels and between (**a**) and (**b**). **c** Histograms showing the distribution of the normalized squared contrast $\mathcal{C}^2/\langle\mathcal{C}^2\rangle$ for the set of images from which the example pattern in (**a**) is taken. The three panels correspond to the three integration lengths shown in (**b**)

to this observable. Figure 3.15 shows examples for the contrast extraction (a, b) and the resulting contrast distributions (c).

For two uncoupled gases in thermal equilibrium this full distribution function of interference contrasts was first discussed in Refs. [65, 66] and applied experimentally in Ref. [67]. It also serve as a sensitive probe for non-equilibrium dynamics as demonstrated in Refs. [42, 68–70]. However, for the analysis presented in this thesis they only serve as a means of thermometry. For that we fit the distribution of the squared contrasts normalized by their mean $\mathcal{C}^2/\langle\mathcal{C}^2\rangle$ with equilibrium predictions obtained from simulated data. This quantity is experimentally more robust as unwanted effects that reduce the contrast, such as the angle of the imaging light with respect to the fringe orientation, can be ignored. A stochastic process is used to obtain single realizations of the phase field in quadratic approximation [61], with the finite imaging resolution considered by smearing the $e^{-i\varphi(z)}$ factor whose integration over the considered length scale gives the contrast (see Ref. [31], Eq. (3.12)).[29] Examples for fits of the measured contrast distributions with these simulated results are presented in Fig. 3.16. For a detailed discussion on the influences of the detection process on the contrast distributions the reader is referred to Ref. [31].

Thermometry via the contrast distributions provides a key advantage over fitting the equilibrium phase correlation function, as discussed in Sect. 3.3.4. Due to the

[29]The soundness of this phenomenological consideration of the imaging resolution was confirmed by numerical simulations if the imaging process. Numerical integration was performed using the novel technique found in Ref. [71].

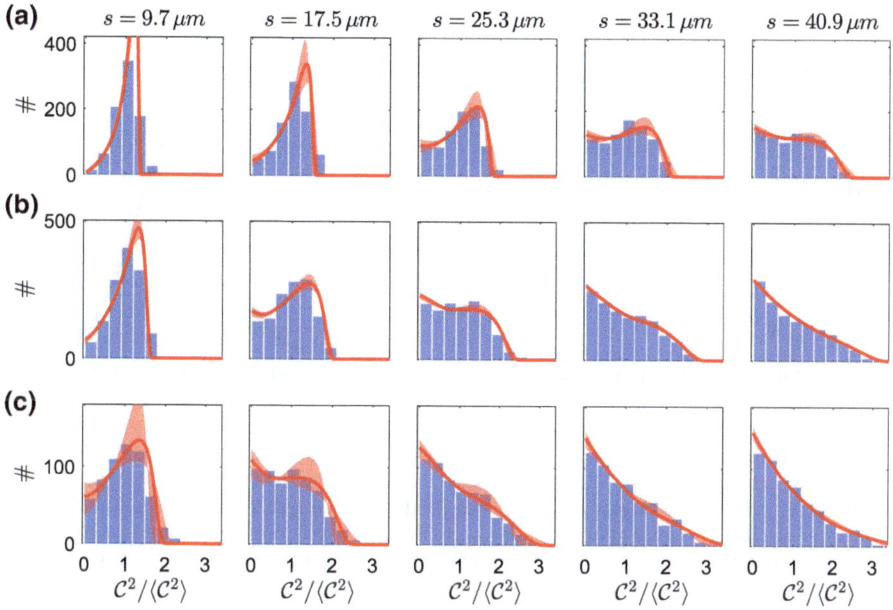

Fig. 3.16 Examples of contrast distribution fits. Histograms showing the distribution of the normalized squared measured contrast $C^2/\langle C^2\rangle$ for five different integration length s compared to the corresponding fits (red lines). Three measurements fitted with different temperatures are presented: **a** $T = 13$ nK (37 nK, 50 nK), **b** $T = 72$ nK (66 nK, 77 nK) and **c** $T = 100$ nK (79 nK, 123 nK). The values in the brackets give the confidence intervals obtained from a reduced χ^2 fit, further visualized by the red shaded areas around the fit curves

inherent integration and the higher stability of the contrast extraction for low fringe visibility the method is not limited to low temperatures. It therefore allows to estimate the temperature of the relative degrees of freedom in regimes where the phase extraction is not reliable anymore. However, as we can see from the fits in Fig. 3.16c the confidence intervals at high temperatures become quite large. This is linked to the fact that λ_T is proportional to $1/T$ (see Sect. 2.1.3), meaning that the relative change in correlation length gets smaller for higher temperatures. Further, as λ_T gets small compared to the used integration lengths the contrast distributions obtained for different integration lengths become more and more similar.

3.3.6 *Balance Measurement*

An important auxiliary tool for experiments performed in the double well potential is the measurement of the population imbalance between the two clouds. For this we separate the clouds after their release from the trap and measure the number of particles in each well individually. The separation is achieved by ramping down the

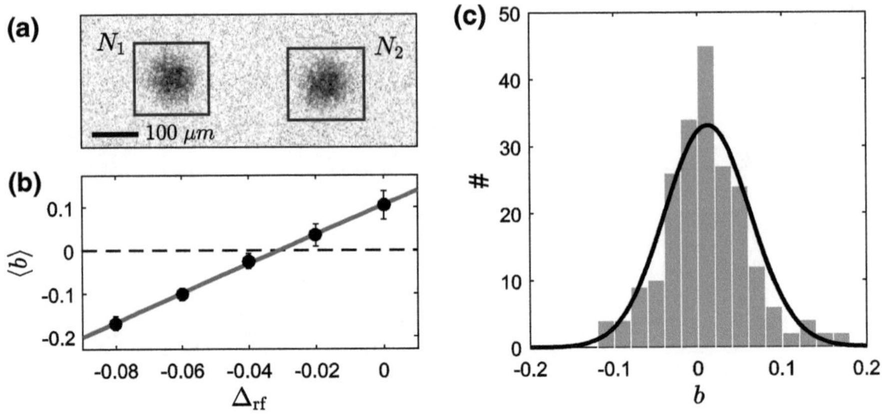

Fig. 3.17 Balance measurement. **a** Atomic density distribution of clouds separated when the double well trap is turned off. The boxes indicate the area integrated to obtain the respective atom numbers $N_{1,2}$. **b** Imbalance between the clouds for different current imbalances between to the rf wires [see fn. 30]. The solid gray line shows a linear fit. **c** Typical histogram of measured balances for a fixed double well configuration. The distribution shows a mean balance of $\langle b \rangle = 0.011$ and a standard deviation of $\sigma_b = 0.051$. The black line shows a Gaussian distribution with exactly these parameters, agreeing well with the data

amplitude of the dressing rf field in 0.1 ms, just before the static trap is turned off. This brings the two clouds in the bare harmonic confinement, off-centered with respect to the minimum of the trap. In the short time before the switch off they acquire enough momentum to separate in time-of-flight (Fig. 3.17a).

We define the imbalance of the clouds as

$$b = \frac{N_1 - N_2}{N_1 + N_2},$$

where $N_{1,2}$ is the number of particles in each of the wells, respectively. For balanced clouds $\langle b \rangle = 0$, which can be achieved by tuning the relative amplitude of the currents in the rf wires (Fig. 3.17b)[30]. However, even when balanced on average the balance distribution exhibits an inevitably spread. In the case of coupled quasi-condensates this spread results from intrinsic thermal fluctuations and technical fluctuations. After the decoupling process used to initiate the recurring dephasing dynamics discussed in Chap. 5 we observe a typical standard deviation of $\sigma_b \simeq 0.05$. When splitting a single well in two thermal fluctuations are absent but atomic shot noise needs to be considered [42].

[30]The parameter $\Delta_{\rm rf}$ gives the relative difference of the squared peak-to-peak amplitude of the applied voltages, translating into a current amplitude imbalance. Due to possible differences in resistance $\Delta_{\rm rf} = 0$ does not equate balanced rf currents.

References

1. Foot CJ (2005) Atomic physics. Oxford University Press, Oxford
2. Metcalf HJ, van der Straten P (1999) Laser Cooling and Trapping. Springer, New York
3. Dalibard J, Cohen-Tannoudji C (1989) Laser cooling below the Doppler limit by polarization gradients: simple theoretical models. J Opt Soc Am B 6:2023
4. Wing WH (1984) On neutral particle trapping in quasistatic electromagnetic fields. Prog. Quantum Electron 8:181–199
5. Steck DA (2015) Rubidium 87 D line data. http://steck.us/alkalidata
6. Ketterle W, Durfee D, Stamper-Kurn D (1999) Making, probing and understanding Bose-Einstein condensates. Proc Int Sch Phys Enrico Fermi 140:67–176
7. Reichel J (2011) Trapping and manipulating atoms on chips. In Reichel J, Vuletić V (eds) Atom chips, Chap. 2. Wiley-VCH, Weinheim, Germany, pp 33–60
8. Reichel J, Vuletić V (eds) (2011) Atom chips. Wiley-VCH, Weinheim
9. Folman R, Krüger P, Schmiedmayer J, Denschlag J, Henkel C (2002) Microscopic atom optics: from wires to an atom chip. Adv At Mol Opt Phys 48:263–356
10. Fortágh J, Zimmermann C (2007) Magnetic microtraps for ultracold atoms. Rev Mod Phys 79:235–289
11. Lesanovsky I et al (2006) Adiabatic radio-frequency potentials for the coherent manipulation of matter waves. Phys Rev A 73:033619
12. Perrin H (2013) Les Houches lectures on adiabatic potentials. http://www-lpl.univ-paris13.fr/bec/bec/Team_Helene.htm
13. Zobay O, Garraway BM (2004) Atom trapping and two-dimensional Bose-Einstein condensates in field-induced adiabatic potentials. Phys Rev A 69:023605
14. Lesanovsky I, Hofferberth S, Schmiedmayer J, Schmelcher P (2006) Manipulation of ultracold atoms in dressed adiabatic radio-frequency potentials. Phys Rev A 74:033619
15. Göbel, M (2008) Low dimensional traps for Bose-Fermi mixtures, Ph.D. thesis, Ruperto-Carola University of Heidelberg
16. Hofferberth S, Fischer B, Schumm T, Schmiedmayer J, Lesanovsky I (2007) Ultracold atoms in radio-frequency dressed potentials beyond the rotating-wave approximation. Phys Rev A 76:013401
17. Grimm R, Weidemüller M, Ovchinnikov YB (2000) Optical dipole traps for neutral atoms. Adv At Mol Opt Phys 42:95–170
18. Bloch I (2005) Ultracold quantum gases in optical lattices. Nat Phys 1:23–30
19. Meyrath TP, Schreck F, Hanssen JL, Chuu C-S, Raizen MG (2005) Bose-Einstein condensate in a box. Phys Rev A 71:041604
20. Gaunt AL, Schmidutz TF, Gotlibovych I, Smith RP, Hadzibabic Z (2013) Bose-Einstein condensation of atoms in a uniform potential. Phys Rev Lett 110:200406
21. Corman L et al (2014) Quench-induced supercurrents in an annular Bose Gas. Phys Rev Lett 113:135302
22. Barredo D, Lienhard V, de Léséleuc S, Lahaye T, Browaeys A (2018) Synthetic three-dimensional atomic structures assembled atom by atom. Nature 561:79–82
23. Hess H et al (1987) Magnetic trapping of spin-polarized atomic hydrogen. Phys Rev Lett 59:672–675
24. Davis KB, Mewes M-O, Joffe MA, Andrews MR, Ketterle W (1995) Evaporative cooling of sodium atoms. Phys Rev Lett 74:5202–5205
25. Petrich W, Anderson MH, Ensher JR, Cornell EA (1995) Stable, tightly confining magnetic trap for evaporative cooling of neutral atoms. Phys Rev Lett 74:3352–3355
26. Davis KB, Mewes MO, Ketterle W (1995) An analytical model for evaporative cooling of atoms. Appl Phys B Laser Opt 60:155–159
27. Luiten OJ, Reynolds MW, Walraven JTM (1996) Kinetic theory of the evaporative cooling of a trapped gas. Phys Rev A 53:381–389
28. Smith DA et al (2011) Absorption imaging of ultracold atoms on atom chips. Opt Express 19:8471–85

29. vom Hagen C (2008) Towards a low-dimensional degenerate Fermi-Fermi-Bose mixture. Ph.D. thesis, Ruperto-Carola University of Heidelberg
30. Kuhnert M (2008) A dual-species two-MOT setup for preparing a Bose-Fermi mixture on an atom chip. Master's thesis, TU Vienna
31. Gring M (2012) Prethermalization in an isolated many body system. Ph.D. thesis, TU Vienna
32. Kuhnert M (2013) Thermalization and prethermalization in an ultracold Bose gas. Ph.D. thesis, TU Vienna
33. Langen T (2013) Non-equilibrium dynamics of one-dimensional Bose gases. Ph.D. thesis, TU Vienna
34. Stix B (2008) A new imaging system for dual-species atom chip experiments. Master's thesis, TU Vienna
35. Schreitl M (2010) Creating and purifying ultracold degenerate gases using hyperfine transitions. Master's thesis, TU Vienna
36. Rauer B (2012) Evaporative cooling of one-dimensional Bose gases. Master's thesis, TU Vienna
37. Reichel J, Hänsel W, Hänsch TW (1999) Atomic micromanipulation with magnetic surface traps. Phys Rev Lett 83:3398–3401
38. Wildermuth S et al (2004) Optimized magneto-optical trap for experiments with ultracold atoms near surfaces. Phys Rev A 69:030901
39. Estève J et al (2004) Role of wire imperfections in micromagnetic traps for atoms. Phys Rev A 70:043629
40. Schumm T et al (2005) Atom chips in the real world: the effects of wire corrugation. Eur Phys J D 32:171–180
41. van Es JJP et al (2010) Box traps on an atom chip for one-dimensional quantum gases. J Phys B At Mol Opt Phys 43:155002
42. Gring M et al (2012) Relaxation and prethermalization in an isolated quantum system. Science 337:1318–1322
43. Gallego D, Hofferberth S, Schumm T, Krüger P, Schmiedmayer J (2009) Optical lattice on an atom chip. Opt Lett 34:3463
44. Straatsma CJE et al (2015) On-chip optical lattice for cold atom experiments. Opt Lett 40:3368
45. Winkler G (2010) A dipole trap on an atom chip. Master's thesis, TU Vienna
46. Tajik M (2017) Arbitrary one-dimensional optical dipole potentials on an atom chip. Master's thesis, TU Vienna
47. Massignan P, Modugno M (2003) One-dimensional model for the dynamics and expansion of elongated Bose-Einstein condensates. Phys Rev A 67:023614
48. Krüger P, Hofferberth S, Mazets IE, Lesanovsky I, Schmiedmayer J (2010) Weakly interacting Bose gas in the one-dimensional limit. Phys Rev Lett 105:265302
49. Gerbier F (2004) Quasi-1D Bose-Einstein condensates in the dimensional crossover regime. Europhys Lett (EPL) 66:771–777
50. Salasnich L, Parola A, Reatto L (2002) Effective wave equations for the dynamics of cigar-shaped and disk-shaped Bose condensates. Phys Rev A 65:043614
51. Pitaevskii L, Stringari S (2003) Bose-Einstein condensation. Clarendon Press, Oxford
52. Dettmer S et al (2001) Observation of phase fluctuations in elongated Bose-Einstein condensates. Phys Rev Lett 87:160406
53. Reinaudi G, Lahaye T, Wang Z, Guéry-Odelin D (2007) Strong saturation absorption imaging of dense clouds of ultracold atoms. Opt Lett 32:3143
54. Schweigler T (2019) Correlations and dynamics of tunnel-coupled one-dimensional Bose gases. Ph.D. thesis, TU Vienna
55. Marte A (2003) Feshbach-Resonanzen bei Stößen ultrakalter Rubidiumatome. Ph.D. thesis, Technical University of Munich
56. Langen T (2013) Comment on probing phase fluctuations in a 2D degenerate Bose gas by free expansion. Phys Rev Lett 111:2012–2013
57. Ockeloen CF, Tauschinsky AF, Spreeuw RJC, Whitlock S (2010) Detection of small atom numbers through image processing. Phys Rev A 82:061606

58. Imambekov A et al (2009) Density ripples in expanding low-dimensional gases as a probe of correlations. Phys Rev A 80:033604
59. Manz S et al (2010) Two-point density correlations of quasicondensates in free expansion. Phys Rev A 81:031610
60. Rohringer W (2014) Dynamics of one-dimensional Bose gases in time-dependent traps. Ph.D. thesis, TU Vienna
61. Stimming H-P, Mauser NJ, Schmiedmayer J, Mazets IE (2010) Fluctuations and stochastic processes in one-dimensional many-body quantum systems. Phys Rev Lett 105:015301
62. Langen T, Geiger R, Kuhnert M, Rauer B, Schmiedmayer J (2013) Local emergence of thermal correlations in an isolated quantum many-body system. Nat Phys 9:640–643
63. Langen T et al (2015) Experimental observation of a generalized Gibbs ensemble. Science 348:207–211
64. Schweigler T et al (2017) Experimental characterization of a quantum many-body system via higher-order correlations. Nature 545:323–326
65. Polkovnikov A, Altman E, Demler E (2006) Interference between independent fluctuating condensates. Proc Nat Acad Sci USA 103:6125–9
66. Gritsev V, Altman E, Demler E, Polkovnikov A (2006) Full quantum distribution of contrast in interference experiments between interacting one-dimensional Bose liquids. Nat Phys 2:705–709
67. Hofferberth S et al (2008) Probing quantum and thermal noise in an interacting many-body system. Nat Phys 4:489–495
68. Kuhnert M et al (2013) Multimode dynamics and emergence of a characteristic length scale in a one-dimensional quantum system. Phys Rev Lett 110:090405
69. Adu Smith D et al (2013) Prethermalization revealed by the relaxation dynamics of full distribution functions. New J Phys 15:075011
70. Kitagawa T, Imambekov A, Schmiedmayer J, Demler E (2011) The dynamics and prethermalization of one-dimensional quantum systems probed through the full distributions of quantum noise. New J Phys 13:073018
71. Tai MM (1994) A mathematical model for the determination of total area under glucose tolerance and other metabolic curves. Diabet Care 17:152–154

Chapter 4
Cooling Through Uniform Loss

Cooling through a controlled loss of particles, in the form of evaporative cooling, is a foundational technique in the field of cold atoms. In nearly all experimental setups it is an essential tool to cool atoms below the limits of laser cooling and to reach degeneracy. Its key ingredients are the energy selective removal of particles paired with an efficient thermalization process (see Sect. 3.1.5). Also in experiments with 1d Bose gases evaporative cooling serves as the final cooling stage. While in optical lattice setups the 1d confinement is usually ramped up after the cooling stopped [1–3], in magnetic micro-traps the 1d regime is generally reached through the evaporation process. On atom chips, the final evaporation segment thus happens when the gas is effectively 1d. However, as discussed in Chap. 2 thermalization is strongly suppressed in the 1d regime, rendering standard evaporative cooling ineffective. One would therefore assume that such cooling ceases to work. Nevertheless, in experiments, cooling under the extraction of particles is observed and cold samples deep in the 1d regime can be obtained [4, 5]. In this chapter we will examine this phenomenon and present a novel cooling mechanism that neither relies on thermalization nor on an energy selective outcoupling of particles [6, 7]. This mechanism fills an important gap in the knowledge of state preparation and the limits of cooling in atom chip setups working with 1d Bose gases.

The chapter starts with a detailed description of the experimental findings, presenting measurements of the temperature evolution under evaporation for different experimental conditions. Further, after investigating the outcoupling process, we discuss the theory of the novel cooling mechanism and compare its predictions to the experimental observations. A final discussion interprets the findings while focusing on the puzzling absence of atomic shot noise effects in the measured cooling sequences.

B. Rauer, *Non-Equilibrium Dynamics Beyond Dephasing*,
Springer Theses, https://doi.org/10.1007/978-3-030-18236-6_4

4.1 Experimental Observation

As mentioned in Sect. 3.1.5, evaporation in our setup is facilitated by coupling atoms to untrapped spin states through rf or microwave transitions. In the standard cooling protocol, the final stage of the experimental cycle consists of an exponential rf frequency ramp reaching a final value a few kHz above the trap bottom, i.e. the Zeeman energy splitting in the center of the trap. This frequency is then held for 100–400 ms. During this final cooling stage the initially thermal gas of atoms first condenses into an elongated 3d condensate and later transitions into the 1d regime. This transition occurs when both the thermal energy $k_B T$ and the chemical potential μ drop below the energy spacing of the transverse motional states $\hbar\omega_\perp$ (see Sect. 2.2). Past this point, we observe a further reduction of temperature under a continuing loss of particles although thermalizing two-body collisions are expected to freeze out fast (see Sect. 2.2.3).

In order to quantify the cooling in the 1d regime we measure the temperature of a single quasi-condensate during the extraction of particles or for different final particle numbers. Figure 4.1 shows such cooling trajectories, with the temperatures inferred from the density ripple thermometry method described in Sect. 3.3.3. The trajectories are obtained from different measurement procedures detailed in the figure caption. All procedures start from a single 1d quasi-condensate keeping the evaporative cooling sequence before and over the 3d $-$ 1d transition the same in order to limit the observed cooling effects to the 1d regime. In all cases we observe a clear cooling effect that reaches down to $k_B T = 0.14\,\hbar\omega_\perp$ for the coldest measurement. This corresponds to $k_B T = 0.32\,\mu$ showing that temperatures well below the chemical potential can easily be reached.

As thermalizing collisions are strongly suppressed in the 1d regime, one would expect the dissipation of particles to drive the system into a non-equilibrium state. However, throughout the whole cooling process the measured g_2 functions stay close to their thermal form, indicating that the state of the gas does not deviate much from thermal equilibrium (Fig. 4.1 inset).[1] This is also observed in the double well potential when analyzing the correlations in the relative phase field between two independent gases cooled by the same procedure (see Fig. 3.14a). Even for coupled gases the evaporation process is shown to prepare equilibrium states when performed slow enough [9]. As phase correlations have been shown to act as sensitive probes for states with non-thermal mode occupations [10] these observations strongly indicate that the evaporation process indeed approximately preserves thermal equilibrium.

Another remarkable observation is that, although the cooling procedures differ significantly between the measurements presented in Fig. 4.1, they all seem to show a linear decrease of temperature with the total number of particles.

[1]Only when the particle number drops further, approaching the lowest densities we can resolve, deviations from the thermal form of the g_2 start to appear. This behavior was documented in Ref. [8] and is not understood up to now.

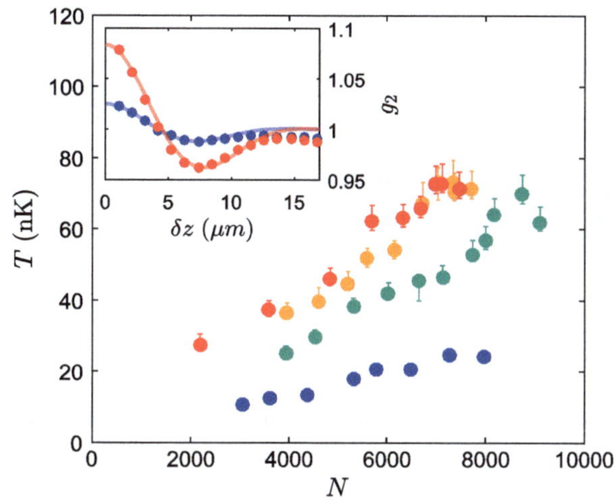

Fig. 4.1 **Cooling measurements**. Temperature over the total number of particle for different cooling procedures. The green and yellow data are recorded at different points in time during a 25 ms long continuous rf induced outcoupling of particles from a harmonic trap with $\omega_\perp = 2\pi \cdot 2.1$ kHz and $\omega_z = 2\pi \cdot 11$ Hz. They only differ in their initial temperature and particle number. The blue data is obtained in the same way but over a timescale of 90 ms in a weakly dressed trap. This trap is still a single well but slightly anharmonic in the transverse directions, with $\omega_\perp \simeq 2\pi \cdot 1.6$ kHz and $\omega_z = 2\pi \cdot 8$ Hz. For the red data atoms are outcoupled from the same weakly dressed trap via microwave transitions to the anti-trapped $|F = 1, m_F = 1\rangle$ state. There, the data points are all taken at the same point in time but the preceding outcoupling rate was changed. The inset shows the g_2 functions of the density ripples corresponding to the coldest data points of the blue and red data. They are compared to simulated data corresponding to the extracted temperatures (solid lines) showing good agreement. All error bars give the 68% confidence interval obtained from a bootstrap (Appendix C)

4.2 Outcoupling Process

To understand the observed cooling behavior we first turn to the outcoupling mechanism and investigate its peculiarities in the 1d regime. In a 3d system undergoing standard evaporative cooling, an energy selective extraction is desired, primarily outcoupling the most energetic particles (see Sect. 3.1.5). However, this does not seem to be the case in the measurements presented in Fig. 4.1, since without efficient thermalization such an outcoupling would drive the system far from equilibrium.

For a direct measurement of the rf induced extraction of particles in 1d we perform a pulsed outcoupling from a quasi-condensate and observe both the outcoupled atoms and the source cloud in a single image (Fig. 4.2). Comparing their spatial distribution we find no indications for energy selectivity. Along the 1d axis the outcoupled atoms show the same average spatial distribution (Fig. 4.2b) and further also seem to exhibit the same density ripple patterns (Fig. 4.2a). This points to a nearly homogeneous outcoupling mechanism. However, note that in this measurement the rf amplitude was about 4 times larger than in the standard final cooling segment in order to

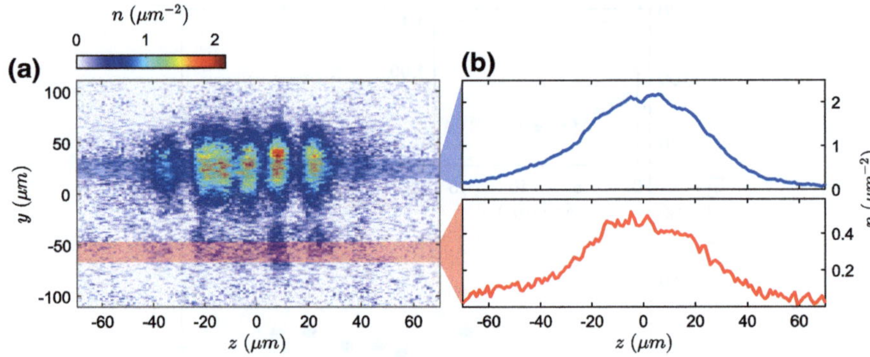

Fig. 4.2 Pulsed outcoupling measurement. **a** Absorption image of single realization of an expanded quasi-condensate subjected to a 2 ms rf pulse 2.5 ms before the trap is turned off. The pulse couples a small amount of atoms to untrapped states such that they leave the trap earlier and show up below the source cloud on the image. Comparing the pattern in the outcoupled cloud with the ripples in the source cloud shows clear similarities, indicating that there is no strong energy selectivity of the outcoupling process. **b** Longitudinal density distribution of the source cloud (upper panel, blue) and the outcoupled cloud (lower panel, red) averaged over many realizations. To avoid any overlap only the blue and red shaded regions indicated in (**a**) are used for the averages. The agreement between the two indicates that atoms are outcoupled homogeneously. The measurement was performed in a double well potential meaning that the transverse image in (**a**) shows the summed signal from two clouds which, however, should not affect the conclusions

obtain a large enough signal from the outcoupled atoms. High rf amplidtudes lead to dressing and thereby to a deformation of the trap such that a direct comparison to the measurements presented in Fig. 4.1 can not be made.

Nevertheless, also calculations of the outcoupling rate Γ support the hypothesis of a near-homogeneous extraction of particles. This rate is proportional to the overlap between the trapped and the outcoupled wave function. Let us first consider the most simplified model of rf outcoupling: the coupling of a single particle in the ground state of a harmonic potential to the states of a free particle experiencing no potential. There, Γ is given by the overlap of the trapped Gaussian wave function (Eq. 2.31) and the free plane wave solutions of the continuum. Figure 4.3a shows Γ in dependence of $\delta = \omega_{\text{rf}} - \omega_0$, the detuning of the rf photon energy ω_{rf} from the transition energy between the atomic states ω_0. For $\delta < 0$, the final continuum state has a positive kinetic energy and Γ is proportional to the overlap of the corresponding plane wave with the trapped Gaussian, which decays exponentially. For $\delta > 0$, no final states are available and Γ drops to zero.

The sharp feature at $\delta = 0$ (Fig. 4.3a) results in an ideal energy selectivity for this simplified outcoupling situation. If the rf frequency is set to slightly positive δ, atoms in longitudinal states (which were not considered in the calculation) with an energy higher than δ get into resonance with the rf while atoms in lower energy states are not affected at all. However, experimentally, when outcoupling atoms to states that are insensitive to the magnetic trap, they still experience gravity which significantly influences their structure. The solutions to the Schrödinger equation in

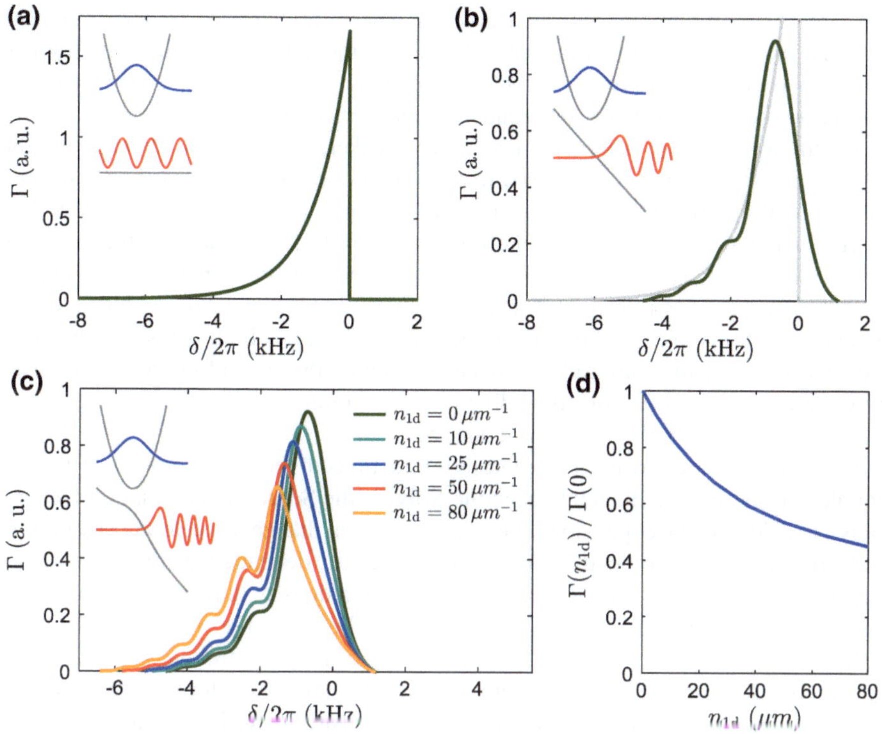

Fig. 4.3 Outcoupling rate. a Transition rate Γ of an atom in the single particle ground state of an harmonic confinement with $\omega_\perp = 2\pi \cdot 2.1\,\text{kHz}$ (blue) into a continuum of untrapped plane waves (red), given in arbitrary units over the rf detuning δ. The dark gray lines in the inset schematics represent the trap potentials. **b** Transition rate for the same configuration but under consideration of gravity. The light gray line shows the curve from (**a**) for comparison. **c** Density dependence of Γ arising when considering the repulsion the source cloud exerts on an outcoupled atom. This calculation also takes the transverse broadening into account. **d** $\Gamma(n_{1d})$ obtained from the calculation in (**c**) for $\delta/2\pi = 0.5\,\text{kHz}$ showing only a small dependence for higher densities

a linear potential are Airy functions (Fig. 4.3b inset). In this case the easiest way to obtain $\Gamma(\delta)$ is to numerically evaluate the spatio-temporal overlap of the trapped wave function $\phi(x, y)$ and the falling and expanding untrapped state $\phi_{\text{out}}(x, y; t)$ [11]

$$\Gamma(\delta) \propto \int_0^\infty \mathrm{d}t\, e^{-i\delta t} \int_{-\infty}^{\infty} \mathrm{d}x\mathrm{d}y\; \phi^*(x, y)\, \phi_{\text{out}}(x, y; t)\,. \tag{4.1}$$

Figure 4.3b shows the resulting rate for atoms trapped in a harmonic potential with $\omega_\perp = 2\pi \cdot 2\,\text{kHz}$ coupled to states that get accelerated away from the trap with $g = 9.8\,\text{m/s}^2$. Due to the linear gravitational potential there are final states available for all δ, leading to a smearing of $\Gamma(\delta)$ that extends to positive δ. The shift of the maximum of $\Gamma(\delta)$ can be understood from the shape of the Airy function. The first and largest Airy maximum is located at a point where the energy of the linear potential lies a bit below the energy of the state. Therefore, the maximal overlap with trapped Gaussian is obtained not for $\delta = 0$ but for slightly negative detunings. Also the minor oscillatory behavior of $\Gamma(\delta)$ for negative δ stems from the Airy shape.

As the longitudinal confinement is much weaker, the motional states along the 1d axis are spaced dense compared to the transverse energy scales. For harmonic longitudinal confinements the typical trap frequency for a single gas is $\omega_z \simeq 2\pi \cdot 10\,\text{Hz}$, such that $\omega_\perp/\omega_z \approx 200$. Therefore, the gravitational smearing of the steep rise in $\Gamma(\delta)$ seen in Fig. 4.3a happens over a broader energy range than typically occupied in a thermal state. An intuitive way to picture this result is that the timescale on which outcoupled atoms leave the trap is on the order of $1/\omega_\perp$, which is fast compared to the timescales of the longitudinal dynamics. Thus, the typical energy differences between atoms are not resolved leading to a near-homogeneous outcoupling, vastly different from standard evaporative cooling.

Of course, this picture is incomplete as interactions of the source cloud and the outcoupled atoms are not taken into account. Solving the Schrödinger equation numerically [2] to calculate the expansion dynamics of an atom that gets accelerated by gravity and repelled from the trapped cloud we obtain a density dependence of $\Gamma(\delta, n_{1d})$ from numerically integrating Eq. (4.1) (Fig. 4.3c). Such a density dependence can lead to an energy selectivity of the outcoupling process. For example, with an rf detuning set to the slope around $\delta \sim 0$, atoms at the edge of the longitudinal trap where the density is low are expelled at a higher rate than atoms in the center where the density is highest. As atoms with a higher energy reach positions further out at the edge of the trap they have a higher probability of being flipped to an untrapped state. However, when analyzing $\Gamma(n_{1d})$ for rf fields blue detuned from the resonance ($\delta > 0$), corresponding to the typical experimental configuration, we observe only a minor dependence, negligible for the central part of the cloud where density does not vary much. Most atoms therefore experience similar transition rates, which is a situation markedly different from standard evaporative cooling for which outcoupling is generally much more selective.

All these considerations lead to the conclusion that when outcoupling atoms from a 1d quasi-condensate a strong energy selectivity is hard to obtain. Even though the outcoupling rate does show density dependencies, they do not allow to selectively address only the most energetic atoms, especially deep in the 1d regime where the central density is low. This near-homogeneous outcoupling makes the the measurements presented in Fig. 4.1 even more remarkable as both pillars of standard evaporative cooling, efficient rethermalization and energy selectivity, are not present. In the following, a theoretical model will be presented showing that efficient cooling however can still be achieved under these conditions.

---

[2]The radial solver discussed in Appendix B can not be used here as the problem is not radially symmetric due to gravity. Instead a full 2d split-step Fourier spectral algorithm was employed. For the trapped state and the initial state of the expansion dynamics the ansatz of an interaction broadened Gaussian discussed in Sect. 2.2.2 was used.

The full experimental situation is even more complex than the one considered in Fig. 4.3c. As we are working with atoms in $|F = 2, m_F = 2\rangle$, for rf induced outcoupling all five m_F states are coupled simultaneously, of which also $|2, 1\rangle$ is trapped (see Sect. 3.1.1). Atoms leaving the trap via single photon transitions to the untrapped $|2, 0\rangle$ state therefore need to pass through this intermediate trapped state. However, observing the state population[3] while cooling we can observe no accumulation or large population of the $|2, 1\rangle$ state. In the microwave cooling data presented in Fig. 4.1 the atoms are directly flipped to the anti-trapped $|1, 1\rangle$ state, presenting also a slightly different situation than Fig. 4.3c. Also, deformations of the potentials due to a dressing of the atomic states (see Sect. 3.1.3) by the outcoupling rf field are not taken into account in the above calculations.

4.3 Theoretical Model

In this section a novel mechanism to explain the observed cooling in 1d quasi-condensates will be presented [6, 7]. It relies on a reduction in the phonon mode occupations and energies through an non-selective particle loss. This reduction is driven by a scaledown of density fluctuations affecting the occupations and by a lowering of the chemical potential altering the mode energies. If the system is initially in a thermal state and the loss proceeds slower than the mode dynamics, thermal mode occupations are preserved. In contrast to standard evaporative cooling this mechanism neither requires an energy selective extraction of particles nor effective rethermalization as each mode is cooled individually.

In the following we will first discuss the case of a homogeneous system dominated by phononic excitation. The approach taken is inspired by Ref. [12] as it gives a very intuitive picture of the process. Then, we turn to the experimentally relevant case of a harmonically trapped gas and compare the measurements presented in Fig. 4.1 to the model predictions. We further discuss the cooling mechanism for particle-like excitations assuming a Bogolibov dispersion and investigate how well the system can be described by an equilibrium distribution throughout the cooling process. Also, we consider the transverse broadening and how it affects the cooling efficiency. In a final section, we discuss the effects atomic shot noise is expected to have on the process and find a theoretical cooling limit inconsistent with the experimental observations.

4.3.1 Loss Cooling

First, let us consider the effects of loss on a homogeneous 1d quasi-condensate. The loss is assumed to be uniform and non-selective in energy such that each particle

[3]For this we apply a magnetic field gradient to the expanding clouds separating the different m_F states in a Stern-Gerlach type measurement.

experiences the same probability of being extracted from the gas. As our observables are dominated by the low-energy part of the excitation spectrum, we first consider the Luttinger liquid model discussed in Sect. 2.1.2. From earlier discussion we know that the governing Hamiltonian given in Eq. (2.9) can be diagonalized by free phononic modes that exhibit a linear dispersion relation. In thermal equilibrium, for temperatures large compared to the relevant mode energies $k_B T \gg \hbar\omega_k$, we can assume each mode to contain the energy $k_B T$. Within each mode this energy is equally distributed between the phase and density quadrature (see Eq. 2.15) resulting in the expectation values of the squared mode amplitudes to be given by

$$\frac{g_{1d}}{2} \langle |\delta n_k|^2 \rangle = \frac{\hbar^2 k^2 n_{1d}}{2m} \langle |\theta_k|^2 \rangle = \frac{k_B T}{2} . \tag{4.2}$$

With the Hamiltonian being quadratic the fluctuations of δn_k and θ_k are Gaussian, fully determined by the second moments given above. They can be visualized by a Wigner function of the mode as shown in Fig. 4.4a.

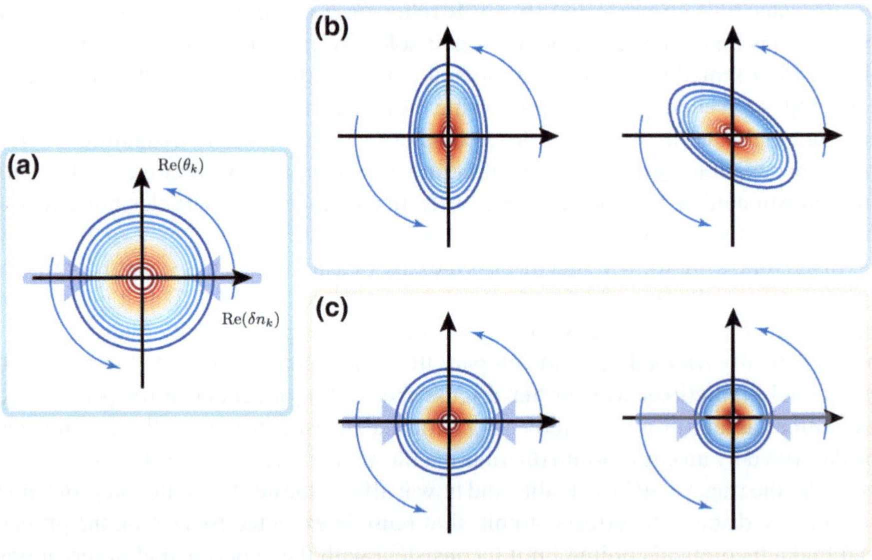

Fig. 4.4 Mode rotation and loss. **a** Contour representation of the Gaussian Wigner function of a single k-mode under equipartition of energy between the density and the phase quadrature. The axis are scaled such that the thermal state given by Eq. (4.2) is radially symmetric. The thick horizontally oriented arrows represent the loss which compresses the distribution in the density quadrature. The thin circularly oriented arrows indicate the rotation of the Wigner function under unitary evolution. **b** An instantaneous loss quench compressed the Wigner function and leads to an asymmetric non-equilibrium distribution. Under unitary time evolution the distribution rotates leading to oscillations in $\langle |\delta n_k|^2 \rangle$ and $\langle |\theta_k|^2 \rangle$. **c** If the extraction of particles is slow compared to the rotation of the mode the Wigner function stays approximately symmetric. The compression is distributed over both quadratures and only the overall width, associated with the effective temperature of the mode, is reduced

Let us now consider the effects of a sudden uniform outcoupling of particles reducing the entire density profile $n(z) = n_{1d} + \delta n(z)$ by a factor α, such that $n(z) \to \alpha\, n(z)$. This reduces both the average density as well as the density fluctuations by this same factor. From the energy balance in Eq. (4.2) we see that such a quench scales down the energy in the density quadrature by α^2 while the energy in the phase quadrature reduces only by α. Therefore, the Wigner function gets compressed along the density quadrature and energy is no longer equally distributed between density and phase fluctuations (Fig. 4.4b). The resulting non-equilibrium state rotates under free evolution leading to phase shifted oscillations of $\langle|\delta n_k|^2\rangle$ and $\langle|\theta_k|^2\rangle$. As the modes do not couple the system does not return to a thermal state in this description.[4]

However, when atoms are extracted continuously at a rate γ slower than the mode rotation the compression is distributed among the quadratures and equipartition is preserved (Fig. 4.4c). Like for a piece of clay on a potters wheel a weak pressure applied along one axis leads to a reduced width in both if the rotation is fast enough. Specifically, the rate needs to fulfill $\gamma \ll 2\omega_k$ as the oscillation between density and phase fluctuations happens twice as fast as the full mode rotation. The change in energy stored in mode k can then be expressed by

$$\delta\mathcal{E}_k = -\,2\gamma\delta t\,\frac{g_{1d}}{2}\,\langle|\delta n_k|^2\rangle \;-\; \gamma\delta t\,\frac{\hbar^2k^2n_{1d}}{2m}\,\langle|\theta_k|^2\rangle\,, \tag{4.3}$$

with the factor of two in front of the first term stemming from the fact that it is proportional to the squared density mode amplitude. Assuming that Eq. (4.2) holds at each point in time we obtain

$$\delta T_k = -\,\frac{3}{2}\gamma\delta t\; T_k, \tag{4.4}$$

for the effective mode temperature. This means that the energy in each mode is independently reduced at the same rate, thereby preserving the initial thermal distribution also without effective rethermalization. The resulting time evolution of the overall temperature is therefore given by

$$T(t) \;=\; T_0\, e^{-\frac{3}{2}\gamma t} \;=\; T_0\left(\frac{n_{1d}(t)}{n_{1d}(0)}\right)^{3/2}. \tag{4.5}$$

This cooling effect stems in part from a reduction of the mode energies and a reduction of the mode occupations. The mode energies are reduced due to the density dependence of the dispersion relation (see Eq. 2.14) while the mode occupations decrease due to the continuous reduction of density fluctuations, visualized in Fig. 4.4. Although being based on quite different premises, this cooling mechanism shows similarities to standard evaporative cooling. In both cases the extraction of particles

[4] See Ref. [13] for an experimental observation of similar dynamics resulting from a quench in the interaction strength.

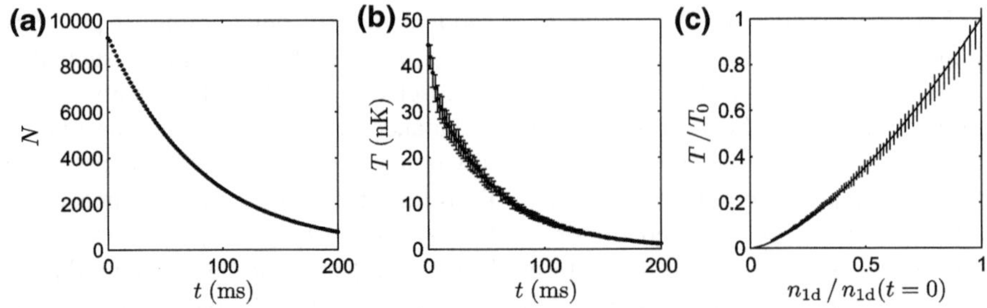

Fig. 4.5 GPE simulations of the cooling. Figures reproduced with permission from [11]. GPE simulation of the cooling dynamics in a homogeneous 1d quasi-condensate subjected to uniform loss. **a** and **b** show the evolution of the total atom number N and the temperature T in time, respectively. **c** shows the development of the temperature rescaled to its initial value in dependence of the rescaled average density. The solid line shows the predicted 3/2 exponent scaling found in Eq. (4.5) agreeing well with the observed temperature evolution

drives the system away from equilibrium while a continuous equilibration mechanism counteracts this process. For evaporative cooling this equilibration stems from thermalizing collisions. For the loss cooling presented here dephasing drives the redistribution of energy within each mode.

Compared to evaporative cooling, this loss cooling is however limited in its efficiency. When considering the Lieb-Liniger phase diagram in Fig. 2.1 we notice that the cooling trajectory defined by Eq. (4.5) moves the system in parallel to the border between the quasi-condensate regime and the ideal Bose gas.[5] This means, although the process can lead to lower mode occupations and therefore bring the system closer to the regime where quantum fluctuations become relevant, it does not bring the gas further into the quasi-condensate regime with respect to the transition to the ideal Bose gas. Nevertheless, the thermal coherence length grows as $\lambda_T(t) \propto e^{\frac{1}{2}\gamma t}$, and equivalently the ratio between the thermal energy and the interaction energy drops $k_\mathrm{B}T/\mu \propto e^{-\frac{1}{2}\gamma t}$.

Figure 4.5 shows how the scaling found in Eq. (4.5) is confirmed by dynamical GPE simulations (see Sect. 2.3.2). Here, the loss was implemented as an additional dissipative term $-i\frac{\gamma}{2}\psi(z)$ and the initial state was sampled by thermally occupying the known plane wave eigenmodes. In Fig. 4.5c, good agreement is found with the 3/2 exponent of Eq. (4.5) indicating that the loss cooling mechanism survives beyond the low-energy approximation.

For the homogeneous case discussed here, the relation given in Eq. (4.5) translates to the same 3/2 scaling of temperature with the total number of particles, as $N \propto n_\mathrm{1d}$. In the following, we will discuss how this mechanism acts in a harmonically trapped system and compare its predictions to the experimentally observed cooling effect.

[5] This is due to $Tn_\mathrm{1d}^{-3/2} \propto t\gamma^{3/2}$ and the quasi-condensate to ideal Bose gas transsition being defined by $t\gamma^{3/2} = 1$. Here, γ and t refer to the parameters of the Lieb-Liniger model defined in Eqs. (2.3) and (2.4), and not the dissipation rate and time!

4.3.2 Trap Effects

The measurements presented in Fig. 4.1 are performed in a harmonic longitudinal trap. To determine the influences of the confinement on the cooling dynamics let us first assuming all atoms to occupy the transverse single particle ground state, neglecting the broadening effects discussed in Sect. 2.2.2. In that case, the equilibrium density profile in Thomas–Fermi approximation is given by an inverse parabola with the central density scaling as $n_{1d}(z = 0) \propto N^{2/3}$, as discussed in Sect. 2.1.2. Therefore, by directly inserting this relation into the scaling found in Eq. (4.5) with n_{1d} replaced by the central density, we find a linear scaling of temperature and atom number

$$\frac{T(t)}{T_0} = \frac{N(t)}{N_0} \,. \tag{4.6}$$

However, this naive translation assumes that the mechanism described in the previous sections is not affected by the inhomogeneous density distribution in the trap, which needs to be shown.

In Thomas-Fermi approximation, the fluctuations in a harmonically confined gas can be decomposed by modes described by Legendre polynomials (see Eq. 2.17). The same argumentation leading to the energy reduction given in Eq. (4.3) applies as well for a homogeneous reduction of an inhomogeneous density profile. But, scaling down the density uniformly drives the profile away from its equilibrium form as the cloud size is not reduced while the atom number decreases (Fig. 4.6a). To reestablish an equilibrium form the profile needs to contract, increasing the central density while reducing the cloud radius R (Fig. 4.6b). This contraction preserves the mode occupations but compresses them spatially leading to a change in the dispersion relation $\omega_j' = \omega_j\, [R(t)/R_0]^{-3/2}$. For the temperature of the gas, proportional to the mode energy, this results in the same scaling with R [14]. Assuming the outcoupling process to be slow enough that the cloud shape adiabatically follows we can separate the incremental temperature changes of the outcoupling δT_{out} and the compression δT_{comp} giving

$$\begin{aligned} \delta T_{\mathrm{out}} &= -\frac{3}{2}\gamma \delta t\, T, \\ \delta T_{\mathrm{comp}} &= -\frac{3}{2}\gamma \frac{\delta R}{R} T = \frac{1}{2}\gamma \delta t\, T \,. \end{aligned}$$

For the second equality of the last line we used the fact that the equilibrium cloud radius scales as $R'/R \propto (N'/N)^{1/3}$ (see Eq. 2.16). Note that, as γ scales down the entire density profile the overall number of particles scales down the same way $\delta N = -\gamma \delta t N$. Adding the two contributions of the temperature change we obtain

$$\delta T = \delta T_{\mathrm{out}} + \delta T_{\mathrm{comp}} = -\gamma \delta t\, T \,,$$

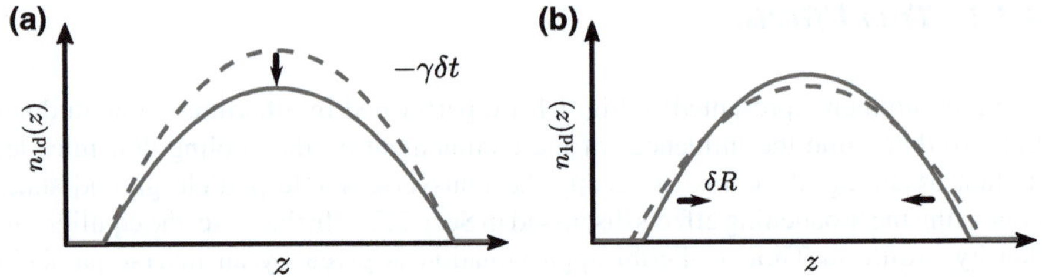

Fig. 4.6 Loss dynamics in an harmonic trap. Schematic depiction of the density profile deformations caused by uniform loss in a harmonically trapped quasi-condensate. **a** The loss leads to a uniform reduction of the local density resulting in a profile larger than the equilibrium size for the now reduced number of particles. **b** Therefore, the system contracts to reach its equilibrium configuration. If the loss rate is slow the profile adiabatically follows its equilibrium form. The derivation in the main text considers infinitesimal changes while here the differences are exaggerated for clarity

Fig. 4.7 Rescaled cooling data. The measurements presented in Fig. 4.1 are rescaled to their respective initial temperatures T_0 and particle numbers N_0. The colors are kept the same

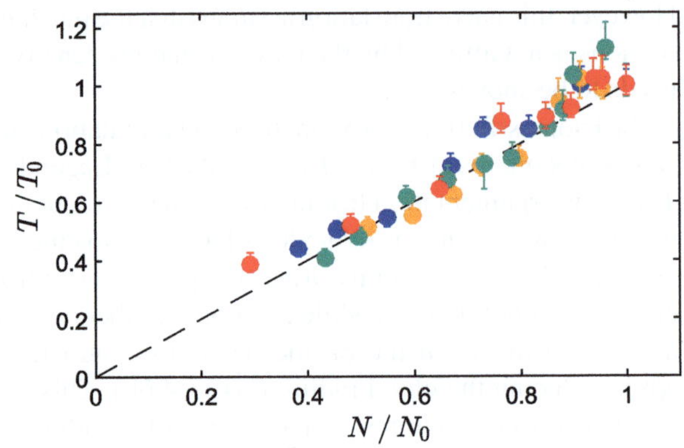

which confirms the linear scaling of Eq. (4.6) found by assuming the result of the homogeneous case applies with n_{1d} replaced by the central density. Note that, in contrast to the homogeneous case, the dispersion relation in the harmonic trap does not depend on density (see Eq. 2.18). Therefore, cooling only stems from a reduction of the mode occupation.

Rescaling the measurements presented in Fig. 4.1 to their respective initial state values T_0 and N_0 we see that the obtained scaling agrees well with the data (Fig. 4.7). When taking a closer look at the loss rates we find that for the fast cooled green and yellow data sets $\gamma \approx 2\pi \cdot 20\,\text{Hz} \simeq 2\omega_{j=1}$, such that the condition of slow outcoupling is barely fulfilled for the lowest mode. For the blue data set $\gamma \approx 2\pi \cdot 12\,\text{Hz}$, which is a bit below $2\omega_{j=1}$. For the red data set the rate is different for each point but the maximum rate can be assumed similar to the one found for the blue data.[6] Nevertheless, the cloud form stays close to its equilibrium shape, indicating that the loss does not substantially drive the system away from equilibrium.

[6] Unfortunately, in this measurement the rate was not independently determined.

As the cooling in the harmonic trap depends on the energy balance between phase and density fluctuations and the shape of the average profile, the transverse broadening will also affect it. As discussed in Sect. 2.2.2, the broadening leads to a distortion of the inverse parabola profile and to a density dependent interaction constant entering Eq. (4.2). Assuming that the simple translation from the homogeneous case applies, as for the inverse parabola profile, we can divide the contribution of the broadening in two parts. First, for the broadened profile the central density does not scale with $N^{2/3}$. When inserting the numerically obtained broadened density scaling into Eq. (4.5) we recover a more efficient cooling (Fig. 4.8). However, the second contribution, that of the modified energy balance, acts in the opposite direction. Considering again the case of a homogeneous gas, replacing g_{1d} in Eq. (2.12) by the density dependent interaction parameter g_n defined through Eq. (2.36), we find the modified incremental temperature change[7]

$$\delta T = -\frac{3}{2}\left(\frac{2+6n_{1d}a_s+5(n_{1d}a_s)^2}{2+7n_{1d}a_s+6(n_{1d}a_s)^2}\right)\gamma\delta t\, T\ . \tag{4.7}$$

In the 1d limit for $n_{1d}a_s \ll 1$ this reduces to Eq. (4.5), but for typical initial values of $n_{1d}a_s \sim 0.6$ the additional prefactor decreases the cooling efficiency. However, adding this effect to the profile distortion described above we find that for typical experimental parameters the overall discrepancy from the linear scaling is too small to observe (Fig. 4.8).

4.3.3 Non-thermal States

The preservation of a thermal equilibrium distribution throughout the loss cooling dynamics discussed above crucially depends on the k-independence of the energy reduction $\delta\mathcal{E}_k$ and the classical fields limit of large mode occupations. Leaving the applicability regime of these approximations we expect the loss to drive the system into a non-equilibrium state as thermalizing collisions are suppressed. In Ref. [12] Johnson et al. calculate the k-dependence of the temperature scaling arising from keeping the quantum pressure term in the second order Hamiltonian given in Eq. (2.8). In momentum space, the Hamiltonian then reads

$$H_{BG} = \sum_{k\neq 0}\left[\left(\frac{\hbar^2k^2}{8mn_{1d}}+\frac{g_{1d}}{2}\right)\delta n_k^2 + \frac{\hbar^2k^2n_{1d}}{2m}\,\theta_k^2\right] + H_0\ .$$

As discussed in Sect. 2.1.2 the eigenmodes of this Hamiltonian follow the Bogoliubov dispersion $\omega_k = \sqrt{E_k(E_k+2\mu)}$ with $E_k = \hbar^2k^2/2m$, going beyond the linear

[7] The given expression is obtained by expanding g_n in powers of $n_{1d}a_s$ and accounting for the different decrease rates for each term when calculating $\delta\mathcal{E}_k$ as in Eq. (4.3). Finally, a closed expression can be found for the resulting series.

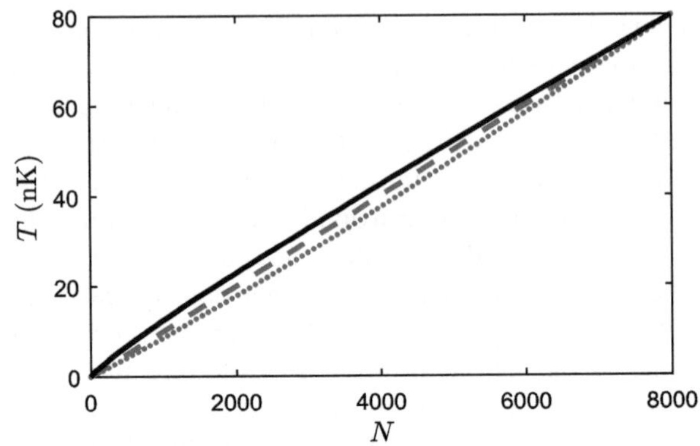

Fig. 4.8 Cooling under transverse broadening in a harmonic trap. Assuming the scaling of temperature with density obtained in Eq. (4.5), but taking into account the distortion of the inverse parabola profile by the transverse broadening, we observe a slightly more efficient cooling (dotted gray line). For comparison the ideal unbroadened linear scaling obtained through $n_{1d} \propto N^{2/3}$ is shown by the dashed gray line. Taking both the broadening effects on the temperature-density scaling given by Eq. (4.7) and the profile distortion into account we see that the two contributions counteract each other and lead to a marginally less effective cooling (solid line). The trap frequencies used here are $\omega_\perp = 2\pi \cdot 2\,\text{kHz}$ and $\omega_z = 2\pi \cdot 10\,\text{Hz}$

phononic regime for high-k modes. Performing the same calculation as in Sect. 4.3.1 we obtain

$$\delta T_k = -\frac{3 + \frac{E_k}{\mu}}{2 + \frac{E_k}{\mu}} \gamma \delta t \; T_k \; . \tag{4.8}$$

Here, δT_k does depend on k, resulting in a different temperature evolution for each mode. For the linear part of the spectrum with $k\,\xi \ll 1$, Eq. (4.8) reduces to Eq. (4.4), leading to the phononic $e^{-\frac{3}{2}\gamma t}$ temperature scaling found previously. For the high-energy particle-like excitations in the limit of $k\,\xi \gg 1$, however, the temperature drops only by a factor $e^{-\gamma t}$. This means the higher modes are not cooled as efficiently and the system is driven into a non-thermal state. As mode occupations are conserved this state can only be described by a generalized form of the standard Gibbs-ensemble [10, 15]. In Ref. [12] it was further shown numerically that within a full GPE description the non-thermal mode occupations obtained from a uniform loss are robust, in contrast to previous observations that mode occupations are not conserved [16, 17].

With our observables dominated by the low-energy, phononic part of the spectrum these non-thermal states elude our experimental observation. However, in a different group they observed discrepancies between the temperatures inferred from *in situ* density fluctuations and the temperatures obtained through fitting the density profile with predictions from the Yang-Yang equation of state [18, 19]. They reported the latter method, which is more sensitive to high-k modes, to give consistently higher temperature then the first. Numerical studies of an harmonically trapped gas described

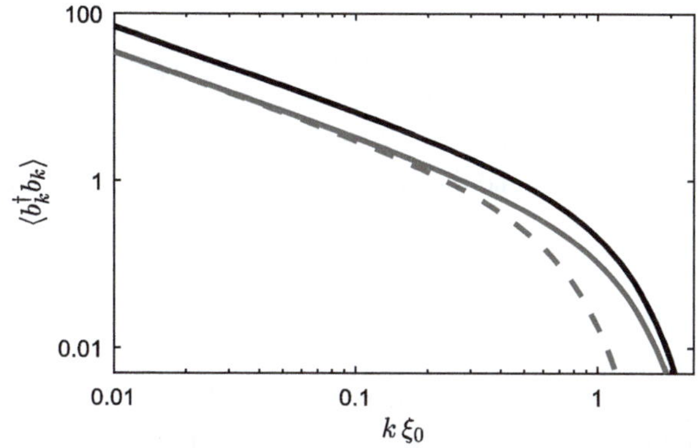

Fig. 4.9 **Loss induced non-thermal state**. Mode occupations $\langle b_k^\dagger b_k \rangle$ under homogeneous loss assuming a Bogoliubov dispersion. The black solid line shows the Bose-Einstein distributed occupations of the initial state with $T_0 = 100\,\text{nK}$, $n_{1\text{d}} = 100\,\mu\text{m}^{-1}$ and $\omega_\perp = 2\pi \cdot 2\,\text{kHz}$. The gray solid lines give the occupations after 50% of the particles were removed by a slow continuous cooling. The dashed gray line shows a thermal distribution at the temperatures obtained from Eq. (4.5) for the cooling in the phononic regime. At higher momenta strong deviations from the thermal state become visible. The k-axis is rescaled by the healing length of the initial state ξ_0

by a GPE with homogeneous loss indicate that these observations can indeed be traced back the non-thermal states induced by the cooling process [20].

Another source for non-thermal occupations is the fact that for high k modes $\langle b_k^\dagger b_k \rangle \gg 1$ does not apply and therefore the assumption that in equilibrium each mode hold the energy $k_\text{B}T$ is not valid anymore. Figure 4.9 shows the non-thermal mode occupations after cooling incorporating both the effect described by Eq. (4.8) and an initial Bose-Einstein distribution. However, note that these curves still result from a classical field approximation which also breaks down at these momenta. A first step towards a quantum mechanical treatment of the loss process and the corresponding cooling effects is described in the following section.

4.3.4 *Shot Noise Corrections*

A crucial detail neglected in all models discussed above is that the loss from an atomic gas is not continuous as atoms are outcoupled one by one. The smooth loss of amplitude of a classical field does not capture the microscopic granularity of this outcoupling process. Moreover, one can picture an atom expelled from the gas at a certain location to leave a 'hole', thereby imparting additional fluctuation on the remaining atom cloud counteracting the cooling process. In Ref. [7] a semi-classical model was derived to describe this noise and quantify its effects on the temperature evolution under uniform loss. In the following we will shortly review this model and compare its predictions to the cooling observed experimentally.

We start again from a homogeneous system well within the 1d regime where the broadening of the transverse wave function can be neglected. Further, neglecting gravity and all interactions of the outcoupled atoms with the trapped cloud we assume the untrapped states to be plane waves. Writing down the coupled equations of motion for the trapped field $\hat{\psi}(z, t)$ and the outcoupled field we can integrate[8] out the latter and obtain [7]

$$i\hbar\,\partial_t\hat{\psi} = -\frac{\hbar^2}{2m}\partial_z^2\hat{\psi} + g_{1\mathrm{d}}\hat{\psi}^\dagger\hat{\psi}\hat{\psi} - i\hbar\frac{\gamma}{2}\hat{\psi} + \hat{\varsigma}\,. \tag{4.9}$$

Here, γ is again the rate of atom loss depending on the overlap between the trapped and the outcoupled states while the last term $\hat{\varsigma}$ represents the shot noise contribution stemming from the discrete nature of the atomic field. Assuming the outcoupled states to be initially empty, occupied only by vacuum fluctuations, and the outcoupling process to be Markovian we find the noise correlations to be linked to the loss rate by $\langle\hat{\varsigma}(z,t)\hat{\varsigma}^\dagger(z,t)\rangle = \hbar\gamma\,\delta(z-z')\,\delta(t-t')$. This white noise properties make sense as the extraction time for a single atom $1/\omega_\perp$ is very short such that the probability of a return to the condensate through reabsorbing a photon is low.

Expressing the field operator in Eq. (4.9) in phase and density fluctuations, again assuming adiabaticity and focusing on the low-energy phononic part of the excitation spectrum in classical field approximation, it can be shown that for a slow loss the ratio between temperature and chemical potential evolves as [7]

$$\frac{k_\mathrm{B}T(t)}{\mu(t)} = \frac{k_\mathrm{B}T_0}{\mu_0}e^{-\frac{1}{2}\gamma t} + \left(1 - e^{-\frac{1}{2}\gamma t}\right)\,. \tag{4.10}$$

The first term on the right hand side gives the same scaling as found for the classical cooling model discussed in Sect. 4.3.1 while the second term stems from the atomic shot noise and does not depend on the initial state. In the limit $t \to \infty$ this term drives the system into a dissipative steady state where the temperature is set by the chemical potential $k_\mathrm{B}T = \mu$ (Fig. 4.10). This state represents the equilibrium between the cooling force arsing from the loss and the heating stemming from its discrete nature. Note however that the final state of the system, for both the classical and the semi-classical model, is the vacuum with all atoms expelled from the trap. The two models only show different routs to arrive at this final state.

This hard limit of the loss cooling mechanism should also affect the experimentally observed cooling. To compare Eq. (4.10) with the measured temperature data we rewrite it to obtain the temperature-density scaling

$$\frac{T(t)}{T_0} = \left(\frac{n_{1\mathrm{d}}(t)}{n_{1\mathrm{d}}(0)}\right)^{3/2} + \frac{\mu_0}{k_\mathrm{B}T_0}\left[\frac{n_{1\mathrm{d}}(t)}{n_{1\mathrm{d}}(0)} - \left(\frac{n_{1\mathrm{d}}(t)}{n_{1\mathrm{d}}(0)}\right)^{3/2}\right]\,. \tag{4.11}$$

[8]Under the assumption that the dynamics of interest happen on timescales larger than $1/\omega_\perp$, the time an outcoupled atom takes to leave the trap.

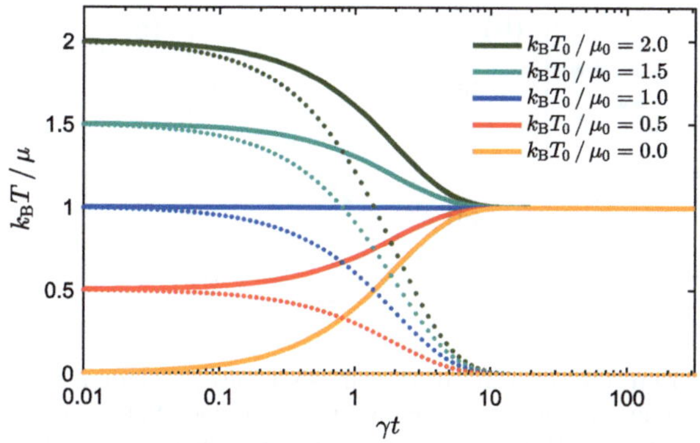

Fig. 4.10 Shot noise induced cooling limit. Thermal energy over the chemical potential during loss cooling in a 1d quasi-condensate. For the classical model discussed in Sect. 4.3.1 (dotted lines) the system evolves towards $k_B T/\mu = 0$ getting more and more coherent. However, taking the shot noise contribution in Eq. (4.10) into account (solid lines) the system runs into a dissipative steady state with $k_B T/\mu = 1$

Again, the first term on the right hand side corresponds to Eq. (4.5) found for the classical case while the second term represents the noise contribution. With the noise term inversely proportional to $k_B T_0/\mu_0$ it only significantly affects the scaling in cases where the initial temperature is low compared to the chemical potential. Figure 4.11 shows a comparison between Eq. (4.11) and three of the data sets presented in Fig. 4.1.[9] As before, we assumed the harmonically trapped system to adiabatically follow the equilibrium density profile such that the central density scales as $N^{2/3}$ and we can replace $n_{1d}(t)/n_{1d}(0)$ by $(N(t)/N_0)^{2/3}$. Effects of the broadening were neglected as we found them not to influence the scaling significantly. Interestingly, a large discrepancy between the shot noise model and the data is observed with the classical linear prediction describing the measured cooling much better. The discrepancy is largest for the data shown in Fig. 4.1a as there the initial state exhibits a temperature of less than half the chemical potential. For the data in Fig. 4.1c $k_B T_0/\mu_0 \sim 1$ such that the differences between the predictions of the classical fields and the shot noise model are small, making it hard to distinguish them experimentally. Note that measuring $k_B T < \mu$ for the initial state alone is no contradiction to the shot noise model as the prior cooling in the 1d-3d crossover might be more efficient.

The coldest data point in Fig. 4.11 reaches down to $k_B T/\mu = 0.32$, far below the expected limit. This is a puzzling result as the introduction of noise through the discrete extraction of particles seems natural. Up to now no satisfying answer resolving this contradiction could be found, however possible explanations will be discussed in the following.

[9] For the fourth data set, the degree of discrepancy with the shot noise prediction lies somewhere between Fig. 4.11b, c but qualitatively shows the same behavior.

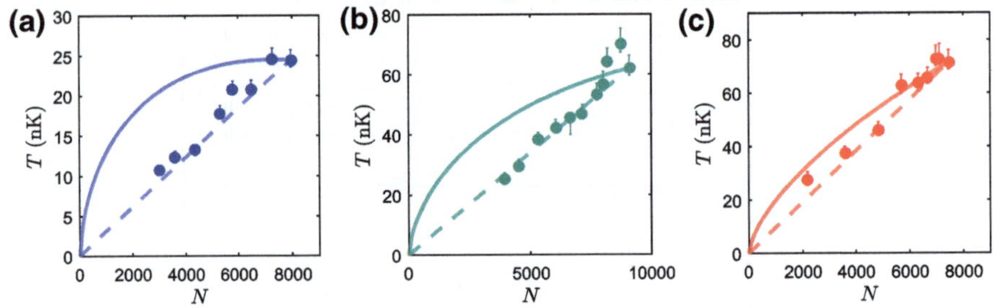

Fig. 4.11 Comparison of the measurements to the shot noise prediction. Measured temperatures for three of the cooling sequences presented in Fig. 4.1, **a** blue, **b** green, **c** red, compared to the classical linear scaling (dashed lines) and the predictions including the shot noise contribution (solid lines). No agreement with the shot noise model can be observed while the classical model describes the cooling behavior very well. The initial states of these measurements show $k_B T_0/\mu_0 = 0.41,\ 0.66$ and 1.25, respectively, when taking the transverse broadening into account. The coldest data point in (**a**) reaches $k_B T/\mu = 0.32$. Note that Eq. (4.11) defining the solid lines does not include the transverse broadening

4.4 Discussion

In this chapter we described a novel cooling mechanism dominant in 1d quasi-condensates subjected to a slow uniform loss of particles. This cooling proceeds without efficient thermalization channels and does not rely on an energy selective outcoupling of particles, setting it apart from standard evaporative cooling. In experiments investigating 1d Bose gases in magnetic traps evaporation is the standard technique to reach degeneracy and the 1d regime. Therefore, this mechanism fills an important gap in the understanding of the state preparation and the limits of cooling.

The results presented in Refs. [6, 7] sparked an interest in the problem and inspired further studies of the cooling mechanism [12, 20–23]. However, the puzzling discrepancy between the experimental observations and the expected shot noise limit described in Sect. 4.3.4 remains unresolved. In Ref. [22] Bouchoule et al. examine the noise induced cooling limit for generic uniform few-body losses in inhomogeneous systems, also taking the transverse broadening into account. They find that under broadening $k_B T$ locks to $mc^2 = n_{1d}\partial_n \mu$ instead of μ and that for a harmonically trapped gas the cooling limit reduces to $k_B T/mc^2 \simeq 0.75$. Yet, the lowest temperatures we observe are still significantly below this limit. Furthermore, the same group recently reported the observation of cooling through recombination induced three-body losses where they see behavior compatible with the predicted cooling limits [23]. However, their initial values of $k_B T_0/mc_0^2$ are already at the predicted limit and they observe no dynamics in that quantity. For regular rf induced outcoupling they observe similar behavior as shown in Fig. 4.11, also reaching temperatures below the expected shot noise induced limit [24].

A possible explanation for these discrepancies might be found in the residual energy-selectivity of the outcoupling process, discussed in Sect. 4.2. Although on the level of single atoms this energy-selectivity should not lead to an additional cooling force due to the lack of thermalizing collisions, it might affect the way excitation are created and annihilated on the level of collective modes [25]. Taking the mode energies into account, outcoupling events that add an excitation to the system should be positively detuned by the energy of that excitation while events that remove an excitation should be negatively detuned. If the rf detuning is set to positive values, on the right hand side of the resonance shown in Fig. 4.3b, c, this would mean that the extraction of excitations happens at a higher rate than their creation. Hence, the amount of energy added to the system would be mitigated, equivalent to a reduced noise contribution. This is related to the fact that an energy-selectivity in the extraction of particles leads to non-Markovian memory effects in the bath and thereby to a coloring of the noise correlations [7]. Yet, for the low-energy modes dominating the experimental observables the difference in detuning for creation and annihilation events is small. It would require a strong energy dependence of the outcoupling rate to obtain significant dentuning differences, which in turn seems unlikely as it would lead to large deviation from thermal equilibrium incompatible with the observations.

A path to further investigate the influence of the outcoupling process would be to compare how cooling proceeds for a truly uniform extraction of particles. Such a uniform outcoupling could for example be realized through two-photon Raman transitions to untrapped states [26]. There, the only requirement for a homogeneous non-selective outcoupling rate would be a uniform intensity pattern of the involved laser beams. Furthermore, outcoupling particles at negative rf or microwave detunings[10] would test the assumption that the specific shape or residual density dependence of the outcoupling rate does not influence the cooling. Tuning the frequency to the other side of the resonance shown in Fig. 4.3c realizes a situation where standard evaporative cooling cannot work, the mechanism presented here however can if the rate is sufficiently uniform.

References

1. Paredes B et al (2004) Tonks–Girardeau gas of ultracold atoms in an optical lattice. Nature 429:277–281
2. Kinoshita T, Wenger T, Weiss DS (2006) A quantum Newton's cradle. Nature 440:900–903
3. Haller E et al (2010) Confinement-induced resonances in low-dimensional quantum systems. Phys Rev Lett 104:153203
4. Hofferberth S et al (2008) Probing quantum and thermal noise in an interacting many-body system. Nat Phys 4:489–495
5. Jacqmin T, Armijo J, Berrada T, Kheruntsyan KV, Bouchoule I (2011) Sub-poissonian fluctuations in a 1D Bose gas: from the quantum quasicondensate to the strongly interacting regime. Phys Rev Lett 106:230405
6. Rauer B et al (2016) Cooling of a one-dimensional Bose gas. Phys Rev Lett 116:030402
7. Grišins P, Rauer B, Langen T, Schmiedmayer J, Mazets IE (2016) Degenerate Bose gases with uniform loss. Phys Rev A 93:033634

[10]Microwave transitions would be the better choice in our setup as for rf tuned to negative detungings atoms proliferate in $|F = 2, m_F = 1\rangle$.

8. Rauer B (2012) Evaporative cooling of one-dimensional Bose gases. Master's thesis, TU Vienna
9. Schweigler T et al (2017) Experimental characterization of a quantum many-body system via higher-order correlations. Nature 545:323–326
10. Langen T et al (2015) Experimental observation of a generalized Gibbs ensemble. Science 348:207–211
11. Grišins P (2014) Cooling and thermalization of one-dimensional bosonic systems. Ph.D. thesis, TU Vienna
12. Johnson A, Szigeti SS, Schemmer M, Bouchoule I (2017) Long-lived nonthermal states realized by atom losses in one-dimensional quasicondensates. Phys Rev A 96:013623
13. Schemmer M, Johnson A, Bouchoule I (2018) Monitoring squeezed collective modes of a one-dimensional Bose gas after an interaction quench using density-ripple analysis. Phys Rev A 98:043604
14. Rohringer W et al (2015) Non-equilibrium scale invariance and shortcuts to adiabaticity in a one-dimensional Bose gas. Sci Rep 5:9820
15. Rigol M, Dunjko V, Yurovsky V, Olshanii M (2007) Relaxation in a completely integrable many-body quantum system: an ab initio study of the dynamics of the highly excited states of 1D lattice hard-core bosons. Phys Rev Lett 98:050405
16. Grisins P, Mazets IE (2011) Thermalization in a one-dimensional integrable system. Phys Rev A 84:053635
17. Kulkarni M, Lamacraft A (2013) Finite-temperature dynamical structure factor of the one-dimensional Bose gas: from the Gross-Pitaevskii equation to the Kardar-Parisi-Zhang universality class of dynamical critical phenomena. Phys Rev A 88:021603
18. Fang B, Carleo G, Johnson A, Bouchoule I (2014) Quench-induced breathing mode of one-dimensional Bose gases. Phys Rev Lett 113:035301
19. Fang B, Johnson A, Roscilde T, Bouchoule I (2016) Momentum-space correlations of a one-dimensional Bose gas. Phys Rev Lett 116:050402
20. Johnson A (2016) One-dimensional Bose gases on an atom chip: correlations in momentum space and theoretical investigation of loss-induced cooling. Ph.D. thesis, University of Paris-Saclay
21. Schemmer M, Johnson A, Photopoulos R, Bouchoule I (2017) Monte Carlo wave-function description of losses in a one-dimensional Bose gas and cooling to the ground state by quantum feedback. Phys Rev A 95:043641
22. Bouchoule I, Schemmer M, Henkel C (2018) Cooling phonon modes of a Bose condensate with uniform few body losses. SciPost Phys 5:043
23. Schemmer M, Bouchoule I (2018) Cooling a Bose gas by three-body losses. Phys Rev Lett 121:200401
24. Schemmer M, Institut d'Optique, Université Paris Saclay, France, maximilian.schemmer@institutoptique.fr (Personal communication)
25. Carusotto I, BEC Center and Dipartimento di Fisica, Università di Trento, Italy, iacopo.carusotto@unitn.it (Personal communication)
26. Hagley EW et al (1999) A well-collimated quasi-continuous atom laser. Science 283:1706–1709

Chapter 5
Recurrences

Each finite and isolated physical system undergoing dynamics will return arbitrarily close to its initial configuration after a certain amount of time. This fundamental consequence of Hamiltonian dynamics, illustrated in Fig. 5.1, was first formulated by Poincaré in 1890 [1]. Later the recurrence theorem sparked the seminal debate between Boltzmann and Zermelo [2, 3] and stimulated a statistical interpretation of the second law of thermodynamics. For quantum mechanics a similar theorem can be proven stating that the wave function of an isolated system with a discrete energy spectrum returns arbitrarily close to its starting point after a finite time [4–6].

As the recurrence timescales for generic many-body systems are astronomically long, easily exceeding the age of the universe, most of the discussions of this phenomena focus on fundamental aspects of thermodynamics and statistical mechanics. Experimental illustrations of the periodicity of the microscopic dynamics are mostly limited to small systems. The revival dynamics in the interaction of a single atom with a cavity field [7, 8] or the periodic collapse and return of coherence between a few atoms localized in isolated sites of an optical lattice [9, 10] are some of the most prominent examples. Yet, partial revivals of the initial state were observed also for more complex many-body systems, in small 1d spin chains realized in ion traps [11] or Rydberg dressed atoms in optical lattices [12].

For larger systems several problems arise. On the one hand the increasing complexity of the eigenstate spectrum leads to exceedingly long recurrence times prohibiting their experimental observation. On the other hand it becomes exponentially difficult to measure the entire state of the system leaving the return of observables as the only viable recurrence criteria. However, the essential dynamical features of large many-body systems can often be described by much simpler effective field theories [13]. In these effective models the complexity of many interacting constituents can reduced to a few collective modes. If these modes are engineered to be almost commensurately spaced in energy a recurrence of the observables dominated by these modes can be realized even in systems of a few thousand particles.

B. Rauer, *Non-Equilibrium Dynamics Beyond Dephasing*,
Springer Theses, https://doi.org/10.1007/978-3-030-18236-6_5

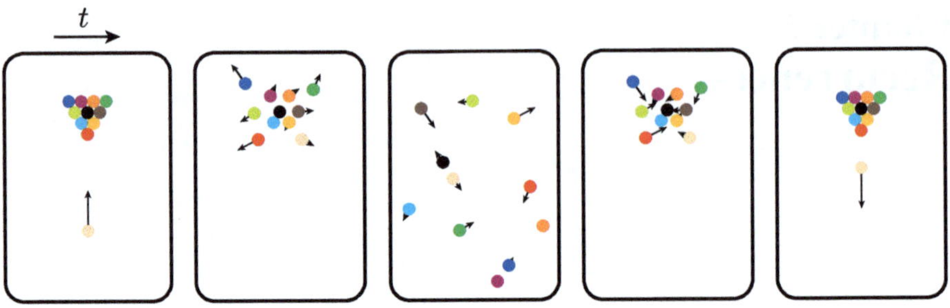

Fig. 5.1 Poincaré recurrence theorem. Illustration of the classical recurrence theorem through the ball movement on an ideal billiard table without friction or pockets. The dynamics starts in the left most panel and recurs shortly after the instance shown in the right most panel

This chapter is dedicated to the observation of recurrences in our model system of two adjacent 1d quasi-condensates [14]. It starts out with a discussion on the role of the dispersion relation in the rephasing dynamics of a multi-mode system and investigates how the most promising conditions for rephasing can be engineered in our system. Following that, we describe the experimental protocol devised to observe recurrences, its implementation and the results obtained. After a general discussion of the measurements a large part of the chapter is dedicated to the recurrence damping and the quest to find its primary sources. In the conclusion we review our findings and discuss them in the general context of quantum simulations.

5.1 Dispersion Relation and Rephasing

The key feature determinating a systems dephasing and rephasing dynamics is the dispersion relation of its excitations. It defines the frequency at which modes evolve relative to each other and therefore after what time their phases realign. In the general case, the greatest common divisor of all frequencies sets the timescale for the return. For an arbitrary spectrum this divisor will be extremely small, leading to tremendously long recurrence times eluding observation. However, if the mode energies are commensurate, following from a simple dispersion relation, the timescale of a mode realignment can be greatly reduced.

For the 1d quasi-condensates realized in our setup the dispersion of excitations is modified by the background density profile $n_{1d}(z)$ which in turn is set by the shape of the longitudinal confinement. In the standard harmonic trap, the mode energies are proportional to $\sqrt{j(j+1)/2}$, as discussed is Sect. 2.1.2. For low j they are therefore incommensurate and do not facilitate a recurrence at experimentally accessible timescales. A detailed discussion of the imperfect rephasing dynamics in the harmonic trap is presented in Sect. 5.5.

In the ideal case of a homogeneous background density the phonon modes are described by plane waves exhibiting an equally spaced dispersion. For the hard wall boundary conditions of an ideal box this leads to Eq. (2.14) and the mode energy spacing $\Delta\omega = c\pi/L$. Independent of the initial conditions, a recurrence of these excitations occurs when all of them performed an integer number of rotations n_j such that the recurrence condition

$$\omega_j t = 2\pi n_j, \tag{5.1}$$

is fulfilled. For the equally spaced mode energies of Eq. (2.14) this happens at $t = 2\pi/\Delta\omega = 2L/c$. At that time the lowest lying mode performed a single turn while the second one performed two turns and so forth ($n_j = j$).

The augmentation of the atom chips magnetic confinement by the dipole trap described in Sect. 3.2.3 allows for the realization of a nearly ideal box potential of variable length. Typical box length lie in the range of few tens of μm and for our parameter regime the speed of sound c lies between 1 and 2 μm/ms (see Fig. 2.6). This brings the recurrence time well within the experimentally accessible time frame. In the following we will describe how the discussed rephasing can be realized and observed in our setup.

5.2 Experimental Implementation

Although the rephasing condition discussed in the previous section applies for any initial state and shows up in the density fluctuations as well as the phase fluctuations, its measurement demands a more sophisticated approach. First, as the fluctuations in a single gas are not easily accessible, a pair of quasi-condensates trapped in a transverse double well potential is employed. In such a configuration matter-wave interferometry provides direct access to the relative phase field between the two gases, as discussed in Sect. 3.3.4. For two uncoupled gases the low-energy dynamics of these relative fluctuations is governed by a Luttinger liquid Hamiltonian, just like the fluctuations in a single well (see Sect. 2.1.3). Up to second order they completely decouple from the fluctuations common to both gases. All considerations on the rephasing of phonon modes discussed above therefore also apply for these relative degrees of freedom. If the low-energy approximations holds sufficiently long such that interactions between the phonon modes can be neglected over the corresponding time scales the observation of recurrences should be feasible.

Another crucial ingredient to experimentally measure a recurrence is the ability to create a distinct initial state. In principle, any state evolving under the Hamiltonian of Eq. (2.9), ontop a sufficiently homogeneous background density $n_{1\mathrm{d}}$, will return to its origin. However, for the experimental implementation it is important to choose a state that can be reproduced and that provides a large signal at its returns. As our main observable is the relative phase field $\varphi(z)$, the latter can be achieved by creating a state with long range phase correlations. This can be realized either by splitting a single gas in two, as shown in Refs. [15, 16] or by suddenly decoupling two strongly coupled gases. For the measurements presented in this thesis we opted for the decoupling scheme as it provides a cleaner initial state with less spurious excitations. The intricacies of the splitting process are discussed in more detail in Sect. 5.2.4.

In the following, the measurement procedure, the obtained results and their interpretation will be presented.

5.2.1 Measurement Scheme

To prepare the strongly coupled initial state we directly cool the atoms into a double well potential with a low enough barrier height to allow for phase locking. The final stage of evaporative cooling is performed much slower than the cooling into a single well, employed for the measurements presented in Chap. 4. The rf frequency is first ramped down exponentially over $\sim$30 kHz in 470 ms and then held constant for 400 ms. This slow cooling sequence assures that no spurious solitons are excited in the process. It is adapted from Ref. [13] where it was shown that the final state created is indeed consistent with thermal equilibrium predictions of the sine-Gordon model.

At the beginning of this final cooling sequence the optical dipole trap amplitude is ramped up in 200 ms. The height of the added box walls is $\sim$1.3 kHz and stays unchanged after its initialization. It is only turned off together with the magnetic trap when the gases are released for measurement.

The decoupling of the wells, initiating the non-equilibrium dynamics, is performed immediately after the cooling rf field is ramped down by increasing the rf current in the dressing wires from $\sim$20 to 24 mA. In about 2 ms the barrier between the wells is raised and their separation increased, completely shutting off the tunneling coupling between the quasi-condensates. As it is not possible to pin down the exact point of decoupling, all times given in the following are with respect to the middle of this ramp.[1] The fact that the decoupling is performed right after the cooling rf field is turned off is essential to avoid unwanted dynamics. Apart from the dressing rf field, also the cooling rf field exerts a slight dressing on the atoms which, when turned off, leads to a sudden change in the coupling strength J, triggering unwanted dynamics in the relative degrees of freedom. By starting the decoupling ramp right after the cooling is stopped the effects of this additional quench can be minimized.

After decoupling the evolution of the system is observed through matter-wave interferometry, as described in Sect. 3.3. Repeating this procedure many times for different evolution times t after the quench provides multiple independent realizations of the relative phase field $\varphi(z)$ during the dynamics. From these realizations we are able to calculate the phase correlation function $C(z, z', t)$ (Eq. 2.25). As the effect we intend to measure is global and the system is sufficiently homogeneous in its central region we calculate an averaged translation invariant from of the phase correlations

[1] Since the time scale of the decoupling ramp is much smaller than than the recurrence time scale this choice only has negligible effects on the results presented (see Fig. 5.4c).

$$C(\bar{z}, t) = \frac{1}{n_{\bar{z}}} \sum_{\bar{z}=z-z'} C(z, z', t), \tag{5.2}$$

where $n_{\bar{z}}$ is the number of distinct pairs of points z and z' separated by $\bar{z}$.

5.2.2 *Results*

The phase correlation dynamics resulting from the described quench protocol in a $L = 49\,\mu\text{m}$ long box trap are shown in Fig. 5.2. Initially, the relative phase between the quasi-condensates is locked exhibiting long range correlations over the whole sample, as discussed in Sect. 2.1.3. This represents the equilibrium state of the strongly coupled system shown in Fig. 2.3b. With the decoupling quench at $t = 0$ this state is projected onto the eigenstates of the uncoupled Hamiltonian and immediately start to dephase. During the first ~10 ms we observe this dephasing through a decay of long range correlations and the establishment of a thermal-like state (see green data points in Fig. 5.2c). This completely dephased state, unfolding in a light-cone-like evolution [16], would represent the end of dynamics for a Hamiltonian with a fully incommensurate spectrum.

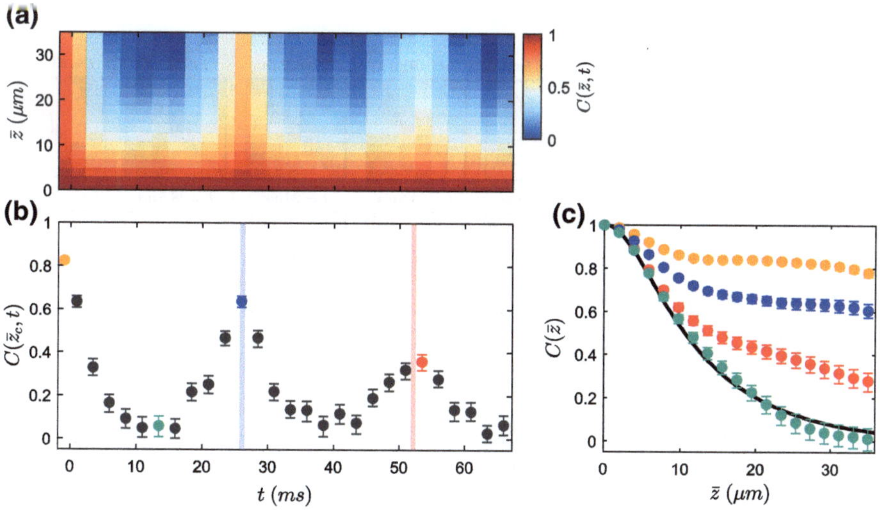

Fig. 5.2 **Phase correlation dynamics after decoupling.** **a** Evolution of the averaged phase correlation function $C(\bar{z}, t)$ (Eq. 5.2) in a box trap with $L = 49\,\mu\text{m}$. **b** Temporal cut of C at $\bar{z}_c = 27.3\,\mu\text{m}$. The error bars give the 68% confidence interval obtained from a bootstrap (Appendix C). The vertical blue and red lines correspond to the times at which the first two recurrences are expected, $t_{\text{rec}} = L/c$ and $2t_{\text{rec}}$ respectively. **c** Spatial phase correlations at different times, corresponding to the colored data points in (**b**)

In the case of the box confinement used here however, the mode energies are commensurate. Therefore, the excitations start to rephase after the initial complete dephasing. In Fig. 5.2a, b this can be seen around $t = 26$ and 52 ms. Contrary to the discussion in Sect. 5.1, the recurrences occur in intervals of

$$t_{\mathrm{rec}} = \frac{L}{c} \, ,$$

indicated by the vertical lines in Fig. 5.2b. The reason for this is that at that time, half way to the full recurrence, the modes rephase to the mirrored initial state $\varphi(z) \rightarrow \varphi(-z)$. While all even modes fully rephase, the uneven modes are half a rotation away from rephasing leading to this inverted state. However, since the observable $C(\bar{z})$ is insensitive to this mirror transformation and the long range correlations in our initial state are not affected by it we see no difference in this half-way recurrence and the full recurrence at $2t_{\mathrm{rec}}$. For the rest of the manuscript we will therefore make no distinction between them.

It is important to note that the observed return of phase correlation is a global effect; it does not depend on the specific locations z and z' at which $C(\bar{z})$ is evaluated. This becomes especially clear when looking at the full two-point phase correlations presented in Fig. 5.3. The recurrence uniformly appears over the whole

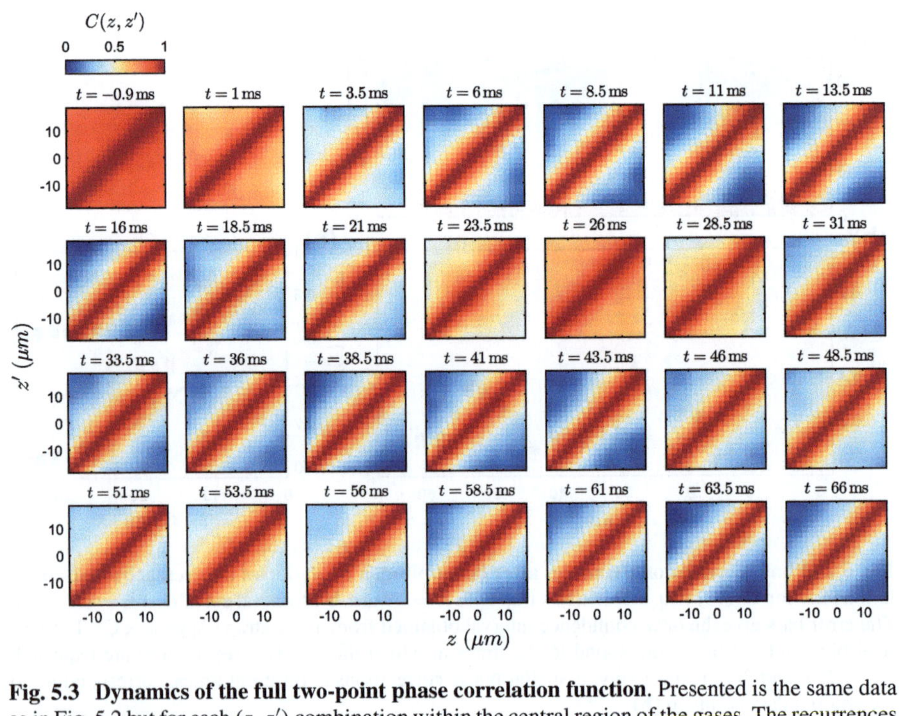

Fig. 5.3 Dynamics of the full two-point phase correlation function. Presented is the same data as in Fig. 5.2 but for each (z, z') combination within the central region of the gases. The recurrences can be seen around $t = 26$ and 52ms

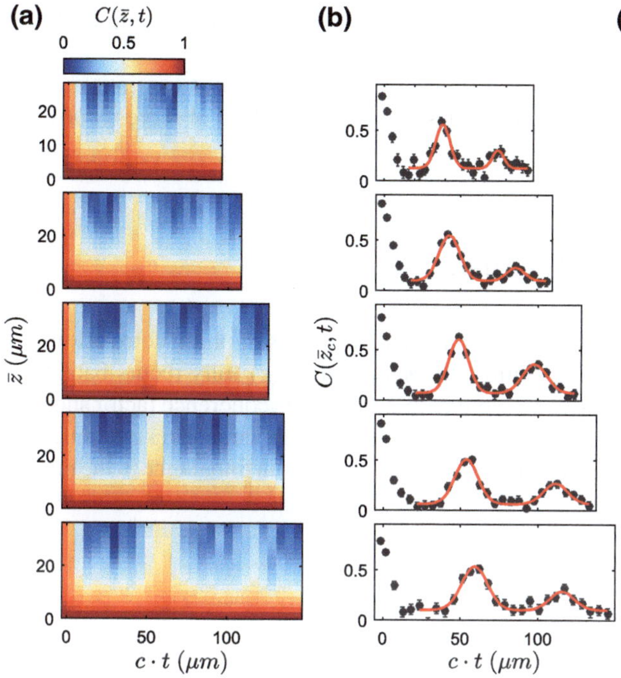

Fig. 5.4 Comparison of different box lengths. a Evolution of the averaged phase correlations $C(\bar{z}, t)$ (Eq. 5.2) in box traps of length $L = 38, 43, 49, 54$, and $60\,\mu$m (top to bottom). The time axis is rescaled by the speed of sound c to make measurements with slightly different central densities comparable. **b** Corresponding temporal cuts of C at $\bar{z}_c = 27.3\,\mu$m. The error bars give the 68% confidence interval obtained from a bootstrap (Appendix C). The red lines show double Gaussian fits to the first two recurrences. **c** Times of the first (blue) and second (red) recurrences extracted from the Gaussian fits. The error bars again give the 68% confidence interval obtained from a bootstrap (Appendix C). The dashed line shows the ideal linear scaling while the green shaded area indicates the uncertainty of the point of decoupling in time

analyzed central region of the clouds with no significant local substructure. This gives an additional justification to the use of the averaged phase correlation function of Eq. (5.2) which corresponds to a mean over the diagonals of the data presented in Fig. 5.3.

To investigate the scaling of the recurrence time with system size we repeated the measurement for different lengths of the box potential. The results are shown in Fig. 5.4a for $L = 38$–$60\,\mu$m. As expected, for larger systems the recurrences are shifted to later times. Fitting a Gaussian to a temporal cut of the phase correlations around the time of the recurrences (Fig. 5.4b) reveals the linear scaling of the recurrence time with L (Fig. 5.4c). Here, the fit is mainly employed to overcome the coarse experimental sampling in time; however, in Sect. 5.3 a justification for its Gaussian form is given. The cut distance $\bar{z}_c = 27.3\,\mu$m is chosen such that it is long enough to observe a high contrast between the recurrences and the dephased state while still being well below the size of the shortest box traps employed.

From Figs. 5.2 and 5.4 we can see that our experimental observations are well described by a rephasing of phonon modes and that we indeed observe recurrences in our system.

5.2.3 Non-recurring Zero Mode

Although we are able to observe high contrast recurrences of phase correlations this does not mean that in each realization the phase field $\varphi(z)$ fully recurs back to its initial configuration. Looking at a different observable, the coherence factor shown in Fig. 5.5a, we see no recurrence. It decays during the initial dephasing dynamics and stays zero from then on. The reason for this is a small random imbalance in the number

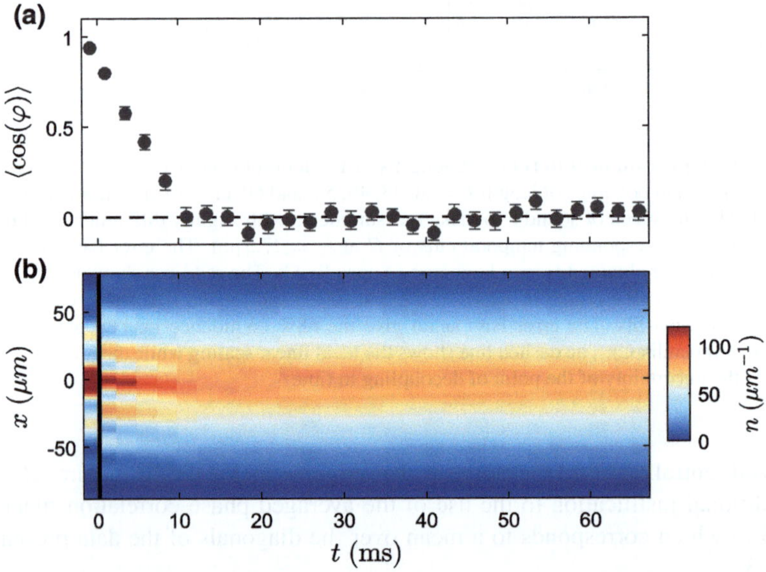

Fig. 5.5 Evolution of the coherence factor and the average interference picture. a Dynamics of the coherence factor $\langle\cos(\varphi)\rangle$ after decoupling, evaluated at the center of the trap at $z = 0$. Due to the near homogeneity of the box trap the behavior is qualitatively the same over the whole sample. The error bars give the 68% confidence interval obtained from a bootstrap (Appendix C). The dashed line is a guide to the eye representing the fully dephased state. **b** Dynamics of the ensemble averaged interference picture, integrated along the z-axis. The black vertical line separates the dynamics after decoupling (right) from the average interference pattern of the initial state (left). As the well separation is increased in the decoupling process, the fringe spacing in the initial state interference pattern is larger. The color map is adjusted to provide maximal contrast for the dynamics after decoupling and therefore locally saturates for the central fringe of the initial state. Due to a small none-zero mean imbalance the central fringe slightly moves to negative x values during the first 10 ms of the dynamics

of particles in each well. This imbalance originates in the thermal fluctuations of the initial state and technical fluctuations. The random particle number difference leads to a running global phase between the wells that accumulates at a different speed for each realization. It can be described as a population of the $k = 0$ mode [17], leading to a dynamics that is not periodic in t_{rec}.

This global dephasing is also reflected in the evolution of the ensemble averaged interference picture, shown in Fig. 5.5b. The straight high contrast fringes of the initial state decay and do not recur. At the time of the recurrence the fringes return to their straight initial state but are shifted by the random global phase, destroying the contrast in an averaged picture.

The phase correlation function $C(\bar{z})$, however, is insensitive to global phase drifts as it only depends on differences of the phase. This illustrates the importance of correlations for our analysis and shows that recurrences or revival effects can be hidden behind a global phase diffusion.

5.2.4 Common Density Dynamics

Up to this point, the discussion of the dynamics induced by the decoupling focused on the relative fluctuations between the two wells. This is due to the relative phase being our main observable but also since in the ideal case, where only the tunnel strength J is changed, the common modes stay unaffected in the initial dynamics (see Sect. 2.1.3). However, as the decoupling process demands a change of the transverse confining potential, spurious effects come into play leading to the common mode density dynamics shown in Fig. 5.6a. At the time of decoupling, density dips are launched from the box walls in both wells and travel through the gases at the speed of sound (dotted green lines). One source for these unwanted excitations is a slight change in the transverse trap frequency of each well, leading to a quench in the interaction strength. As the wells are separated and the barrier is raised they are also tightened from $\omega_\perp \simeq 2\pi \cdot 1.40$ to $2\pi \cdot 1.45$ kHz. This 4% difference leads to a change in the equilibrium density profile that is most prominent at the edges of the system (Fig. 5.6c, d), partly explaining the form of the induced excitations. A second source is a shift of the longitudinal minimum of the dressed state potential during decoupling. This is caused by the chip wire corrugations discussed in Sect. 3.2.2 and explains the asymmetry of the density waves observed in Fig. 5.6a.

The effects of both of these sources are much more severe when splitting a single well in two, rendering this protocol unfeasible in the box trap, though the phase coherent state created would provide an excellent recurrence signal in the relative phase. In previous studies performed in an harmonic longitudinal confinement [15, 16, 18] these common mode excitations only led to slow breathing and dipole oscillation irrelevant for the short time dynamics.

As the dynamics in the longitudinal potential are common to both wells, the observed density waves should have no effect on the recurrences observed in the relative fluctuations. However, since the time the density waves take to traverse the

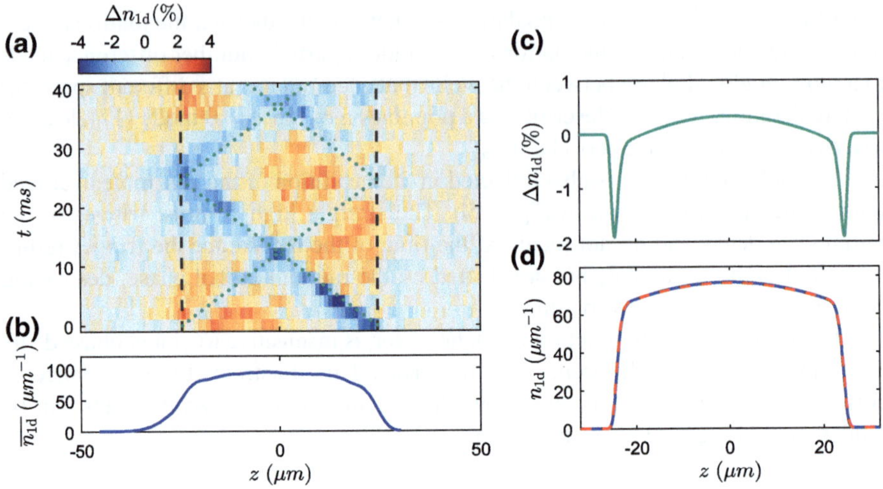

Fig. 5.6 Density waves caused by the decoupling process. a Dynamics of the in situ density profile after decoupling, imaged after 2 ms of free expansion. Plotted is the deviation of the density profile from its temporal mean $\overline{n_{1d}}$ in relation to the maximum density in percent, $\Delta n_{1d}(z) = [n_{1d}(z) - \overline{n_{1d}}(z)]/\max(\overline{n_{1d}}(z))$. The vertical dashed lines indicate the walls of the box. At $t = 0$, one can see small density waves being launched from the walls and traveling through the system. The dotted green lines, sloped at the speed of sound, are a guide to their path. **b** Temporal profile mean. The slight asymmetry of the profile edges is an artifact of an unwanted image shift in the CCD. **c** Difference between the equilibrium profiles before and after decoupling, calculated from an imaginary time evolution of the broadened GPE (see Sect. 2.2.2). **d** Corresponding density profiles before (solid blue line) and after decoupling (dashed red line) illustrating that the induced density waves are only a small perturbation

system naturally coincides with the recurrence time $t_{rec} = L/c$ it has to be ensured that the periodic signal observed in the relative phase correlations can not be influenced by them. To this end, we simulated the dynamics after decoupling in a GPE simulation (see Sect. 5.4.2) including the interaction quench at $t = 0$. The observed common mode excitations look qualitatively similar to Fig. 5.6a but no influence on the relative phase fluctuations could be found on experimentally relevant timescales.

5.3 Theoretical Description

The observation made in Fig. 5.2c, that the dephased state between the recurrences is experimentally indistinguishable from a thermal state is remarkable as it shows how hidden correlations in a seemingly equilibrated system can retain the memory of the initial state. It illustrates how closed quantum systems will equilibrate on average even though their dynamics is inherently recurrent [19]. However, the thermal form of the correlations is still surprising as the mode populations after decoupling result from projecting the strongly coupled initial state onto the modes of the gapless uncoupled Hamiltonian. A priori there is no reason why this should result in a thermal distribution. Further, any non-thermality originating from the quench would persist during this initial dynamics as phonon-phonon interactions should not yet affect the state much. When completely dephased, such non-thermal states are described by a generalized form of the standard Gibbs ensemble [20] and can experimentally be detected [18].

Let us therefore briefly return to the analytic description of the ideal box modes discussed in Sect. 2.1.3 and calculate the post-quench mode occupations by projecting the strongly coupled thermal initial state onto the modes of the gapless uncoupled Hamiltonian. In the new basis this gives

$$\begin{aligned}\langle b_k^\dagger b_k\rangle &= \frac{k_B T_{\text{init}}}{2\epsilon_k}\left(\frac{\mu}{\mu+\hbar J}+\frac{E_k}{E_k+2\hbar J}\right),\\ \langle b_k b_k\rangle = \langle b_k^\dagger b_k^\dagger\rangle &= \frac{k_B T_{\text{init}}}{2\epsilon_k}\left(\frac{\mu}{\mu+\hbar J}-\frac{E_k}{E_k+2\hbar J}\right),\end{aligned} \tag{5.3}$$

in classical field approximation, with T_{init} being the temperature of the coupled thermal state and ϵ_k the linear dispersion of the final Hamiltonian. For a completely dephased state we can calculate the effective mode temperature $k_B T_k = \langle b_k^\dagger b_k\rangle/\epsilon_k$ from the first line in Eq. (5.3). At large k values the dispersion does not change much during the quench such that $T_{k\to\infty} = T_{\text{init}}$, whereas for low k modes the temperature is closer to $T_{k=0} = T_{\text{init}}/2$ (Fig. 5.7a). For intermediate k values T_k lies in between these two limits. Even though the population of modes is clearly non-thermal the finite resolution of the phase measurement (see Sect. 3.3.4) smears the phase correlation function such that it remains still close to a thermal form (Fig. 5.7b, inset). This explains the thermal $C(\bar{z})$ observed between the recurrences in Fig. 5.2c.

From Eq. (5.3) we can further calculate the full dynamics of the phase correlation function. Taking the variance of the phase mode amplitude in Eq. (2.23), for $J = 0$ we obtain

$$\begin{aligned}\langle|\varphi_k|^2\rangle &= \frac{1}{n_{1d}}\frac{\mu}{\epsilon_k}\left[2\langle b_k^\dagger b_k\rangle - \left(\langle b_k b_k\rangle e^{i2\omega_k t} + \text{H.c.}\right)\right],\\ &= \frac{2}{n_{1d}}\frac{\mu}{\epsilon_k}\left[\langle b_k^\dagger b_k\rangle - \langle b_k b_k\rangle\left(1-2\sin^2(\omega_k t)\right)\right],\end{aligned} \tag{5.4}$$

where for the last step we used the fact that for our initial state the anomalous correlators are equal. Inserting this result into Eq. (3.11) and converting the integral to a sum for a finite system, we can calculate the time evolution of the phase correlation function. Note that the factor of two in the exponential time dependence of Eq. (5.4) again illustrates why the obtained signal is periodic in $t_{\text{rec}} = \pi/\Delta\omega$ and not the full recurrence time $2t_{\text{rec}}$.

Expanding Eq. (5.4) for short times around the recurrence at $\omega_k t = \pi n_k$, with n_k being an integer, gives a quadratic time dependence in first order,

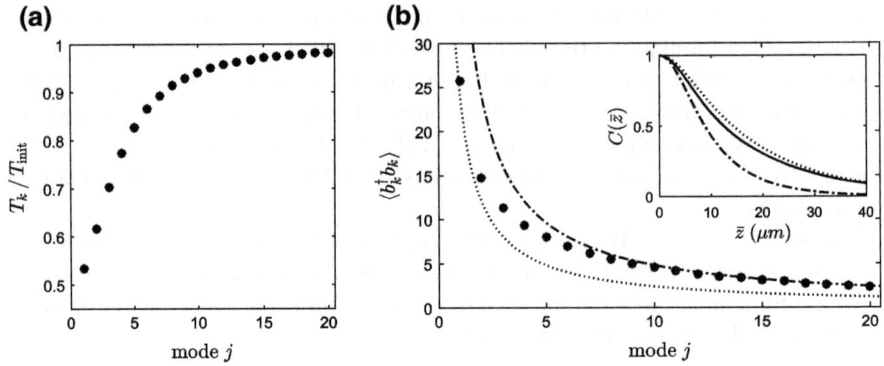

Fig. 5.7 Mode population after decoupling. **a** Effective mode temperatures reflecting the occupations given by Eq. (5.3) for typical experimental parameters ($L = 50\,\mu$m, $n_{1d} = 70\,\mu$m^{-1}, $J = 2\pi \cdot 3$ Hz). **b** Actual occupations for $T_{\mathrm{init}} = 50$ nK and the parameters used in (**a**). The dash-dotted and dotted lines show thermal distributions for T_{init} and $T_{\mathrm{init}}/2$, respectively. The inset shows the corresponding phase correlation functions taking the experimental imaging resolution into account. Here, the solid line represents the phase correlations obtained for a fully dephased state with the occupations shown in (**a**) and (**b**)

$\langle|\varphi_k|^2\rangle = a_k^{(0)} - a_k^{(1)} t^2 + O(t^4)$, with $a_k^{(i)}$ being the k-dependent expansion coefficients. Using this expansion to calculate the temporal evolution of the phase correlation function in the vicinity of the recurrence shows that $C(\bar{z}, t)$ is given by a product of Gaussians with k-dependent widths. As a product of Gaussians is again Gaussian the heuristic fits employed in Fig. 5.4 are well justified.

Calculating the evolution of $C(\bar{z}, t)$ over longer times reveals interesting finite size effects. Additional features appear between the recurrences when analyzing the correlation dynamics between two specific points z, z', without averaging over multiple points that fulfill $\bar{z} = |z - z'|$. The blue line in Fig. 5.8b shows an example of these features for a specific choice of z, z'. At $t_0 = |z - z'|/2c$ the light-cone-like spreading of dephased correlations reaches the distance between the two points and the initial decay of correlations stops; later, at $t_{\mathrm{rec}} = L/c$ the global recurrence occurs. In between, at $t_1 = |z_1 + z_2 - 2z_{\mathrm{B}}|/2c$, where z_{B} is the point of the closest boundary, we see a dip in the correlations which repeats itself at $t_{\mathrm{rec}} - t_1$. This effects is best understood within the picture of correlated quasi-particle pairs that are created locally in the quench and traverse the system during dynamics [21]. Figure 5.8a shows a schematic of their trajectories and explains how their reflection on the boundary of the system leads to the observed features. Of the pair marked in yellow, one quasi-particle travels to the right reaching z' after t_1 while the other first travels to the left, gets reflected off the wall, and then reaches z after t_1. As this folded light-cone brings quasi-particles that originated together to z and z' one would expect a peaked correlation feature. However, due to the Neumann boundary conditions imposed by the hard walls the reflected quasi-particle picks up a π phase shift at the wall, leading to the observed dip. Such boundary induced correlation features are discussed in the

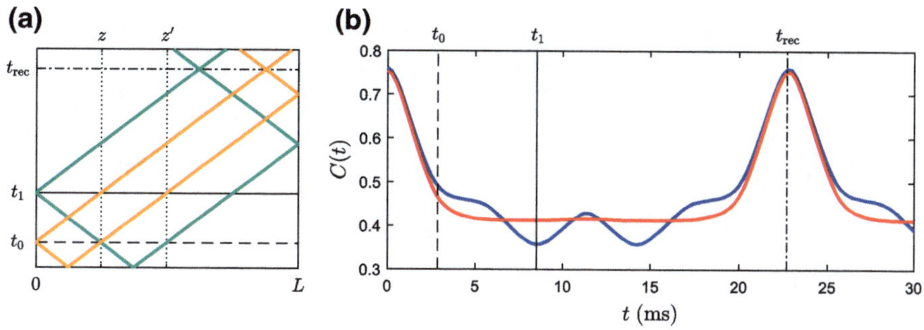

Fig. 5.8 Finite size effects in the phase correlations. **a** Schematics of the quasi-particle propagation after the decoupling quench in an ideal box trap with respect to the two points $z = L/4$ and $z' = L/2$ chosen asymmetrically around the center. The green lines show a quasi-particle pair originating in between z and z' and reaching them after t_0 (dashed black line). The yellow lines indicate a pair of quasi-particles that simultaneously pass the points at t_1 (solid black line) after one reflection. Both pairs return to their mirrored origin ($z \rightarrow -z$) after the recurrence time t_{rec} (dash-dotted black line). **b** Corresponding temporal evolution of the two-point phase correlations $C(z, z', t)$ (blue) compared to the evolution of the averaged phase correlations $C(\bar{z} = |z - z'|, t)$ (red) (used parameters: $L = 50\,\mu\text{m}$, $T = 40\,\text{nK}$, $n = 70\,\mu\text{m}^{-1}$)

literature on post-quench dynamics in finite spin-chains where they are sometimes called 'traversals' [22, 23].

In the experiment, the finite wall steepness and the finite optical resolution generally wash out these dips. This explains why no additional features can be seen in the unaveraged full two-point correlations shown in Fig. 5.3.

5.4 Damping

The dynamics predicted by the quadratic model of the ideal box discussed above is completely periodic in t_{rec}, it entails no damping of the recurrences. Nevertheless, the measurements presented in Figs. 5.2 and 5.4 do show a distinct recurrence damping. The first recurrence does not reach the level of correlation in the initial state and the second one is damped even further. A third recurrence might have been observable in some of these configurations but was not investigated due to the long measurement times necessary.[2] However, in a single high density measurements with coarser time resolution we managed to observe an albeit faint third recurrence, as shown in Fig. 5.9.

[2] For a typical measurement like the one presented in Fig. 5.2 we record at 28 different points during the evolution and additionally perform a balance (Sect. 3.3.6) and density ripple measurement (Sect. 3.3.3). For each data point we repeat the experiment 300 times to obtain the necessary statistics. This demands 9000 runs of the 25 s long experimental cycle leading to a total measurement time of over 60 h.

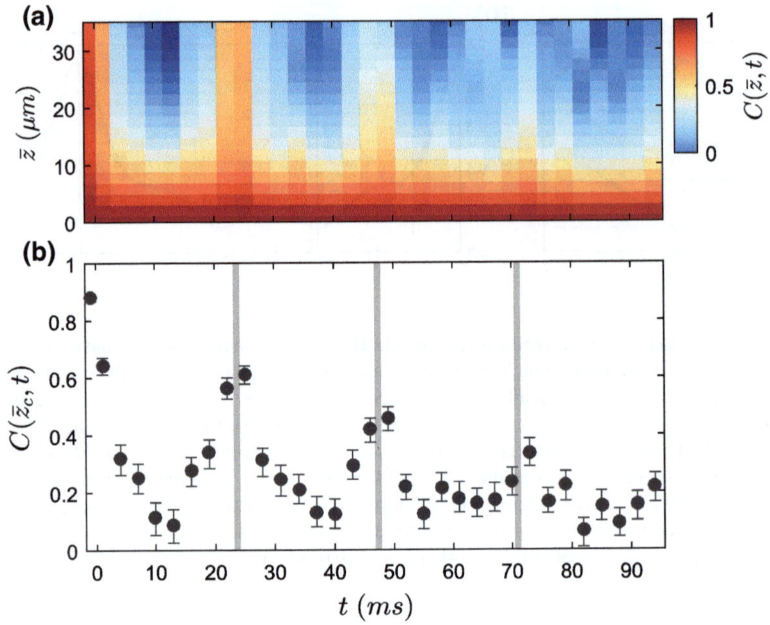

Fig. 5.9 Observation of a third recurrence. **a** Evolution of the averaged phase correlation function $C(\bar{z}, t)$ (Eq. 5.2) in a box trap with $L = 49\,\mu\text{m}$ measured for up to 94 ms. **b** Temporal cut of C at $\bar{z}_c = 27.3\,\mu\text{m}$ with the error bars giving the 68% confidence interval obtained from a bootstrap (Appendix C). The vertical lines indicate the expected recurrence times. Due to the higher central density of $101\,\mu\text{m}^{-1}$, leading to a larger speed of sound, the recurrences occur slightly earlier than in Fig. 5.2

For the ideal system an exactly equally spaced spectrum and a free evolution of phonon modes lead to phase correlations that reproduce themselves till eternity. To investigate why the experimental realization differs from this ideal situation and what the leading damping mechanisms are we increase the initial temperature of the system by intentionally deoptimizing the evaporative cooling process in the Z-trap. The resulting recurrence signals are shown in Fig. 5.10. For higher temperatures the damping increases significantly, until around $T_{\text{eff}} \simeq 100\,\text{nK}$ the recurrences are fully damped. The temperatures given are extracted from the full distribution functions of interference contrasts (see Sect. 3.3.5) around $t = t_{\text{rec}}/2$ in the fully dephased state.[3] They represent an effective measure since the mode occupation of the relative degrees of freedom are not expected to be thermal, as discussed in Sect. 5.3. To isolate the temperature dependence of the damping from its dependence on density

[3] The interference pictures of the four points in time closest to $t = t_{\text{rec}}/2$ are combined in this analysis to obtain a better estimate of the effective temperature. The fits shown in Fig. 3.16 correspond to three of the measurements presented in Fig. 5.10. As discussed in Sect. 3.3.4, the spatial decay of the phase correlations does not allow for a reliable thermometry at such high temperatures.

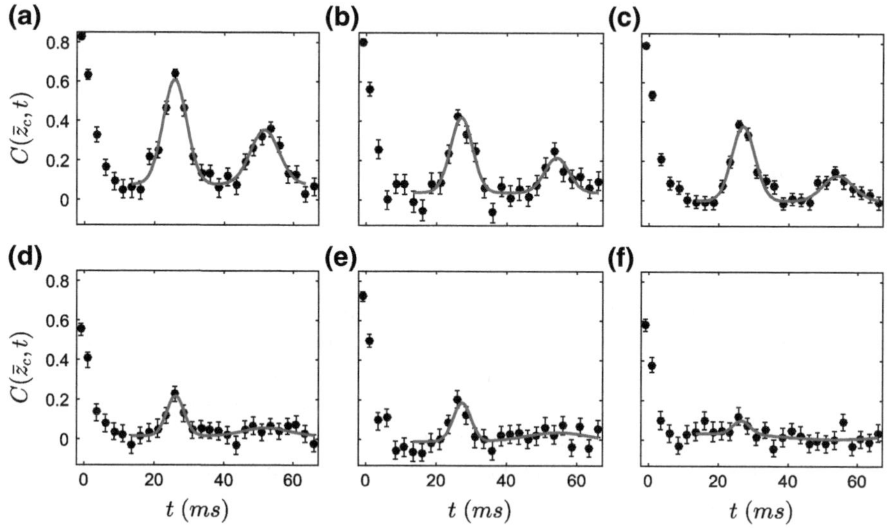

Fig. 5.10 **Recurrence signals for different temperatures**. Temporal cut $C(\bar{z}_c = 27.3\,\mu\text{m}, t)$ for six different effective temperatures **a** $T_{\text{eff}} = 43\,\text{nK}$, **b** 70 nK, **c** 72 nK, **d** 79 nK, **e** 85 nK and **f** 100 nK. The error bars give the 68% confidence interval obtained from a bootstrap (Appendix C). The solid gray lines show the double Gaussian fits employed to extract the recurrence heights (Eq. 5.5)

all measurements presented in Fig. 5.10 are performed at similar central densities of $n_{1\text{d}} = 74$–$79\,\mu\text{m}^{-1}$.

As for the extraction of the recurrence times shown in Fig. 5.4 the fit function used in Fig. 5.10 is a superposition of Gaussians, motivation by the discussion in Sect. 5.3

$$f(t) = C_{\text{base}} + \sum_{i=1}^{2} C_{\text{fit}}^{(i)}\, e^{-\frac{(t - i \cdot t_{\text{rec}})^2}{2\sigma_i^2}} . \tag{5.5}$$

However, to stabilize the fit for the bootstrap procedure (Appendix C) in the cases of low recurrence amplitude, some parameters were fixed. The recurrence time t_{rec} was first separately extracted from an unrestricted fit[4] while the amount of base correlation C_{base} in the fully dephased state (zero line of the Gaussian fit) was obtained from averaging over the data points around $t = t_{\text{rec}}/2,\ 3t_{\text{rec}}/2$ and $5t_{\text{rec}}/2$. This means only $C_{\text{fit}}^{(i)}$ and σ_i in Eq. (5.5) are extracted from the fit. With these restrictions the fitting is stable for all investigated temperature sets and virtually all bootstrap realizations.

There are several sources of damping that contribute to the decay of the recurrence signal. In the following we will first discuss the different damping mechanisms and

[4]Meaning that all parameters in Eq. (5.5) are fitted together. Only for the hottest data set in Fig. 5.10 the recurrence time predicted by the Luttinger liquid model was used due to the low experimental signal.

further try to disentangle their contribution to the observed decay by comparing the experimental results with the predictions of different theoretical models.

5.4.1 Sources

The damping sources can be classified in trivial sources stemming from experimental imperfections and fundamental sources related to many-body processes. A prime source of trivial damping is the spread in atom number of the experimental samples. Each realization inevitably has a slightly different number of particles, which leads to a different speed of sound and therefore to a different recurrence time. The phase correlation dynamics extracted from many realizations is therefore averaging over all these slightly displaced recurrence signals, reducing the observed recurrence height. Typical standard deviations of the number of particles are around 15% of the average. Due to the weak density dependence of the speed of sound for higher densities (see Fig. 2.6) this translates to only a ~5% spread in c for the investigated parameter region. By post-selecting for realizations with similar atom number this effect can be further reduced but never eradicated due to the limited statistics available.[5] Another source of spread is the imbalance in the number of particles between the wells discussed in Sect. 5.2.3. The occupied $k = 0$ mode can also be understood as a difference in speed of sound in the two wells which leads to a coupling between the common and relative degrees of freedom, as discussed in Sect. 2.1.3 and Ref. [24]. With the common degrees of freedom being in a thermal state this coupling also leads to a decay of the recurrence signal.

The second major source of trivial damping is the imperfect box potential. Both the residual longitudinal harmonic trap and the finite steepness of the walls (see Sect. 3.2.3) distort the equally spaced dispersion relation of the perfectly homogeneous system. This leads to a relative detuning of the phonon mode oscillations away from the recurrence condition of Eq. (5.1) and thereby reduces the height of the recurrence signal.

Only a minor role is played by the quadratic part of the Bogoliubov dispersion which also leads to unequal energy spacings in the spectrum. Due to the limited contribution of high energy modes to the dynamics and the finite imaging resolution our observations are always dominated by the linear part of the dispersion relation. Also the total loss of atoms over the measurement time, amounting to ~5%, should only contribute weakly to the damping. This loss stems from three-body recombinations, technical noise and collisions with the background gas [25], all leading to an immediate removal of the involved particles from the trap. Auxiliary measurements in uncoupled equilibrium systems show no significant heating over the investigated time scales, indicating that the loss does not substantially affect the dynamics. As

[5] For the analysis presented in this chapter we generally disregard the lowest and highest ~15% in atom number for each data point.

discussed in Sect. 3.2.3, also the scattering of dipole trap photons can be completely neglected.

Processes that go beyond dephasing or loss are of greater interest. They necessitate scattering channels that allow for the redistribution of momenta or phonon-phonon interactions mediated by higher-order terms in the 1d Hamiltonian. However, as discussed in Sect. 2.2.3, both thermalizing two-body collisions and three-body processes involving virtual excitations of transverse excited states are predicted to be negligible on the observed timescales. Even though for $T = 100\,\text{nK}$, corresponding to the hottest measurement in Fig. 5.10, the rate of thermalizing two-body processes rises to $\Gamma_{2b} = 2\pi \cdot 7\,\text{Hz}$, the resulting thermalization time scale is still much slower than the recurrence damping, which appears to happen in less than 20 ms. This leaves higher-order terms within the 1d description as a possible non-trivial source of damping. As discussed in Sect. 2.1, terms going beyond the quadratic approximation lead to phonon-phonon interactions and the coupling of common and relative degrees of freedom [26–28]. Numerically, these processes can for example be studied by solving the GPE [27, 29, 30], however, their experimental observation is challenging. Due to its fragility, the recurrence signal can pose as a sensitive experimental probe for these otherwise elusive processes. In the following, we will try to unveil the source of the observed damping and show that it can indeed be linked to phonon-phonon interactions.

5.4.2 Theoretical Modeling

To separate the trivial from the non-trivial sources of damping we first simulate the dynamics of the relative degrees of freedom on the level of the inhomogeneous Luttinger liquid, include the damping due to the spread of initial conditions and the imperfect experimental box potential. Then, we numerically solve the GPE for two coupled quasi-condensates, which includes higher-order terms beyond the quadratic approximation.

5.4.2.1 Luttinger Liquid Description

To be able to take the inhomogeneous background density and the transverse broadening effects into account we use the method outlined in Sect. 2.3.1 to numerically calculate the thermal initial state of the strongly coupled system and propagate it with the equations of motion of the uncoupled Hamiltonian. This assumes the quench to $J = 0$ to be instantaneous. In the experiment, the raising of the barrier between the wells proceeds over a time scale of $\sim$2 ms with the actual decoupling happening even faster. Therefore, on the time scales of the recurrence dynamics the tunneling switch off is fast, justifying this simplification.

In Fig. 5.11 the results of the calculations are presented. The effects of a spread in atom number or imbalance and the effects of a finite box wall steepness are investigated separately. Figure 5.11a, c show that the averaging over realizations with different initial conditions indeed leads to a significant damping of the recurrence height. For these curves the phase correlations are calculated for different atom number and imbalance values and are then combined in a weighted average according to the respective distributions shown in Fig. 5.11b, d.[6] In the case of the spread in atom number the single realizations recur perfectly but at different times, leading to an apparent damping in the average. The cut tails of the Gaussian distributions in Fig. 5.11b reflects the experimental post-selection in atom number discussed in Sect. 5.4.1. For the spread in imbalance the individual realizations all recur at the same time but the different speeds of sound in the two gases prevents a perfect rephasing also for a single realization. However, as the difference in the speed of evolution simply adds another low frequency to the problem we find a beating pattern for longer times that leads to a nearly perfect recurrence after $2L/|c_1 - c_2|$, where $c_{1,2}$ are the speeds of sound in the two wells (see Fig. 3 in [24]).[7]

In Fig. 5.11e we see the effects that finite box walls have on the recurrence signal. Apart from the expected damping we also notice a small shift of the recurrence time which is linked to the growing length of the density profile for the smeared out box walls. The density profiles used in these calculations are ground state solutions of the broadened GPE for the potential given in Sect. 3.2.3, numerically obtained from an imaginary time evolution (see Sect. 2.3.2). Their edge shape can be seen in Fig. 5.11f. As the chemical potential is lower than the box wall height the change in the density profile is more pronounced for $z > 0$.

Figure 5.11g shows how all these damping processes combined act for experimentally relevant parameters at different initial temperatures. As the dispersion relation is linear, one might expect that a higher temperature does not affect the damping. However, since the recurrences get narrower with increasing temperature all effects that prevent a perfect rephasing have more severe consequences on the recurrence height, leading to the observed temperature dependence (Fig. 5.11h). In this calculation we also included the residual harmonic potential of $\omega_z = 2\pi \cdot 7\,\mathrm{Hz}$, not included in the case of Fig. 5.11e, f. The harmonic potential counteracts the shift in recurrence time observed in Fig. 5.11e, explaining why we still find good agreement with the ideal recurrence time $t_{\mathrm{rec}} = L/c$ in Fig. 5.4. Although the trend of high temperature states being damped faster is reproduced, the trivial damping sources on the level of the Luttinger liquid description cannot qualitatively reproduce the experimental findings.

The exact value of the initial coupling J does not play a crucial role for these calculations. We therefore fixed the ratio λ_T/l_J to 7 which is in the regime of strong

---

[6] For both distributions the dynamics is calculated for 13 values. They are linearly spaced and lie within $\pm 1\sigma$ for the atom number spread and within $\pm 1.8\sigma$ for the imbalance spread.

[7] A perfect recurrence is found only when the dispersion spacing $\Delta\omega$ is a multiple of the added frequency.

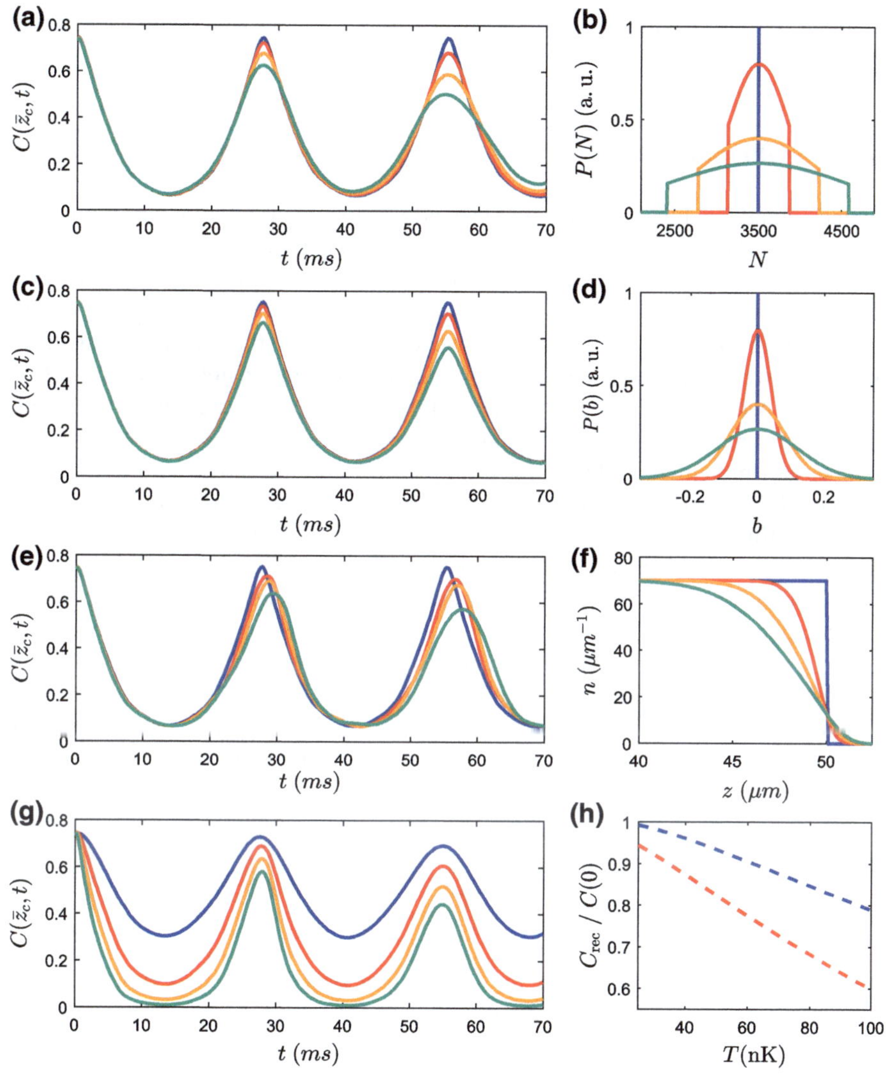

Fig. 5.11 Recurrence damping within the Luttinger liquid model. **a** Dynamics of the averaged phase correlations $C(\bar{z}_c = 27.3\,\mu\text{m}, t)$ in a perfect hard-walled box of $L = 50\,\mu\text{m}$ for different spreads of the total particle number, $\sigma_N/\langle N\rangle = 0, 0.1, 0.2, 0.3$ (blue, red, yellow, green), with the distribution cut at $1\sigma_N$, mimicking the experimental post-selection process, and $\langle N\rangle = 3500$. The temperature of the initial state is set to $T_{\text{in}} = 40\,\text{nK}$. **b** Corresponding particle number distributions. **c, d** The same for different imbalance spreads, $\sigma_b = 0, 0.04, 0.08, 0.12$ (blue, red, yellow, green). **e** Phase correlation dynamics in a box trap with varying wall steepness, $\sigma_{\text{wall}} = 0, 1.2, 2.4, 3.6\,\mu\text{m}$ (blue, red, yellow, green). The total particle number was adjusted such that the central density is $n = 70\,\mu\text{m}^{-1}$ in all cases. **f** Corresponding density profiles at the edge of the system. **g** Phase correlation dynamics for typical experimental parameters ($\sigma_N/\langle N\rangle = 0.15$, $\sigma_b = 0.05$, $\sigma_{\text{wall}} = 1.2\,\mu\text{m}$) calculated for different initial temperature $T_{\text{in}} = 25, 50, 75, 100\,\text{nK}$ (blue, red, yellow, green). **h** Corresponding height of the first (blue) and second (red) recurrence relative to the initial correlations at $t = 0$. For all calculations the transverse broadening was taken into account

coupling. Experimentally we estimate that $J = 2\pi \cdot$ 1–5 Hz which agrees well with this regime.[8]

5.4.2.2 GPE Simulations

To understand the dynamics beyond the quadratic low-energy sector we simulate the system with the GPE, as discussed in Sect. 2.3.2. There, higher order terms mediating scattering between the dephasing and rephasing phonon modes are included. Starting from a thermal initial state of two coupled gases obtained from the SGPE, the tunneling coupling J is instantly quenched to zero and the state is propagated in time with two independent GPEs.[9] As for the experiment, many realizations of the system are needed to obtain converging results for the phase correlations. In the simulations presented here 2500 realizations were used. Fluctuations of the total atom number are implemented through a variation of μ and the spread in the imbalance between the wells is in part intrinsic to the thermal nature of the initial states.[10] As for the Luttinger liquid calculations, the broadening induced density dependence of the interaction term is taken into account as well.

Examples for the obtained phase correlation dynamics are shown in Fig. 5.12. For high temperatures, $T > 90\,\mathrm{nK}$, a strong damping of the recurrences can observed, similar to the experiment. The small 'anti-recurrence' dips visible between the peaks at $t_{\mathrm{rec}}/2$ in Fig. 5.12a, b stem from the fact that here the unaveraged phase correlation function is plotted where z and z' are chose symmetrically around the center of the system. The dips are therefore of the same origin as the features discussed in Fig. 5.8.

5.4.3 *Discussion*

In order to quantitatively compare the discussed models with the experimental observations it is practical to devise a measure for the recurrence height. We chose

$$H_{\mathrm{rec}}^{(i)} = \frac{C_{\mathrm{fit}}^{(i)} - C_{\mathrm{base}}}{C(0) - C_{\mathrm{base}}},$$

where $i = 1, 2$ stands for the first and second recurrence and $C(0)$ is the value of the correlation function before decoupling. The phase correlation values at the time of

---

[8] These values were extracted by comparing the initial state fluctuations with calculated data obtained from a random process simulating the full mean-field sine-Gordon Hamiltonian for a fixed λ_T determined from density ripple thermometry (see Sect. 3.3.3) [13, 31].

[9] As argued above in the case of the Luttinger liquid calculations, an instant quench of the tunneling coupling is a good approximation for the fast experimental decoupling.

[10] Technical sources of imbalance spread are neglected as it is not trivial to implement them in the GPE calculations.

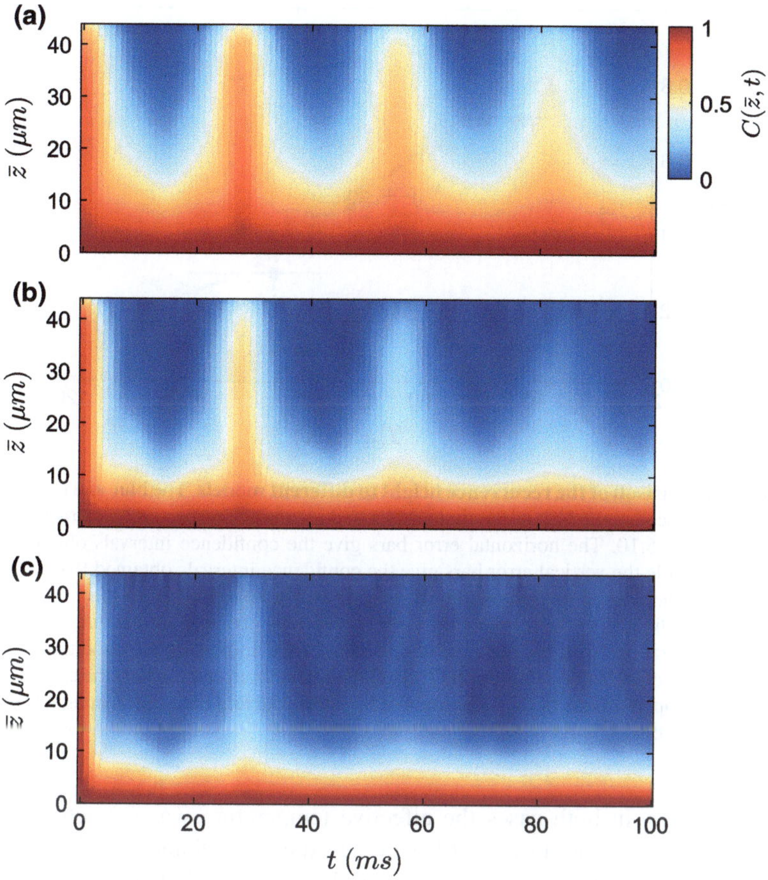

Fig. 5.12 Phase correlations obtained from GPE simulations. Dynamics of $C(\bar{z}_c, t)$ in a realistic box potential of length $L = 49\,\mu\text{m}$ after quenching the coupling to zero for **a** $T_{\text{eff}} = 29\,\text{nK}$, **b** $T_{\text{eff}} = 57\,\text{nK}$ and **c** $T_{\text{eff}} = 103\,\text{nK}$. Here, the phase correlation function is not averaged and z and z' are chose symmetrically around the center of the system. As for the measured phase correlations, T_{eff} is extracted from the dephased state at $t_{\text{rec}}/2$. The simulations were performed by Sebastian Erne and details can be found in Ref. [32]

the recurrences $C_{\text{fit}}^{(i)}$ are extracted from the fits presented in Fig. 5.10. As for these fits, the base correlation value C_{base} is obtained by averaging over the data points of the dephased correlations in between the recurrences. For a perfect return H_{rec} becomes unity while a vanishing recurrence signal gives zero.

Figure 5.13 shows a comparison of the measured recurrence heights, the results obtained from the Luttinger liquid model and the GPE simulations, all evaluated at $\bar{z}_c = 27.3\,\mu\text{m}$. For both models the experimental spread of initial conditions, the form of the experimental potential and the limited imaging resolution are taken into

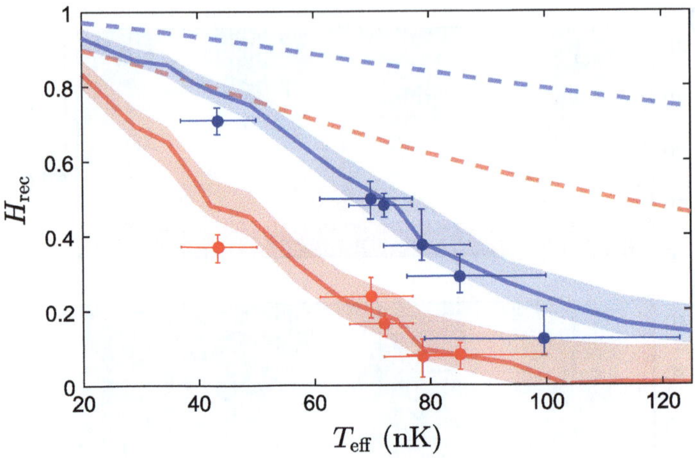

Fig. 5.13 Comparison of the recurrence height to different models. The blue and red data points show the recurrence height H_{rec} of the first and second recurrence respectively, extracted from the fits shown in Fig. 5.10. The horizontal error bars give the confidence intervals obtained from a reduced χ^2 fit while the vertical error bars give the confidence intervals obtained from a bootstrap (Appendix C). The dashed lines show the recurrence heights obtained from the Luttinger liquid model. The solid lines give the result of the GPE simulations obtained from averaging over 2500 independent realizations with the corresponding shaded areas indicating the uncertainty resulting from the limited experimental sample size (1σ deviation). For both calculations the spread of initial conditions, the imperfections of the experimental box potential and the limited imaging resolution were taken into account

account. Further, in both cases the effective temperature was extracted from the spatial decay of correlations at $t_{rec}/2$ as in the experimental analysis.[11]

From the comparison it becomes clear that the trivial damping mechanisms accounted for in the Luttinger liquid model can not explain the damping observed in the experiment. Even for the coldest data point they would predict a much higher recurrence signal. Together with the good agreement reached with the GPE simulations this indicates that genuine phonon-phonon scattering stemming from higher-order terms in the 1d Hamiltonian is the prime source of the damping. This constitutes the first experimental observation of such physics for 1d Bose gases and will form the foundation for several follow-up studies.

Note that the broadened interaction term used in the GPE also partially includes the effects of virtual three-body processes. Just as higher-order terms of the 1d Hamiltonian are included also terms like the one added in Eq. (2.37) are taken into account. However, based on predictions for Γ_{3b} made in Ref. [33] we do not expect these terms to play a significant role in the damping.

[11]This was done by fitting a thermal form (Eq. 2.27) to the spatial decay of correlations. As the Luttinger liquid calculations and the GPE simulations are no plagued by the distortion effects stemming from a finite imaging resolution in the experiment (see Sect. 3.3.4), resorting to an FDF fit was not necessary.

Another important note is that the unreliability of the phase extraction at high temperatures rooted in the finite imaging resolution and discussed in Sect. 3.3.4 does not significantly affect the recurrence signal at large $\bar{z}$. If short range phase jumps are missed in the analysis, $C(\bar{z}_c, t)$ is not affect. This was confirmed by simulations of the imaging process.

5.5 Harmonic Trap

In the beginning of this chapter we briefly mentioned that observing rephasing in a pair of gases trapped in a harmonic longitudinal confinement is not feasible in our setup. Figure 5.14a, showing measured phase correlation dynamics after decoupling in a harmonic trap, validates this statement. The main reason for this qualitatively different behavior is that the phonon modes in the harmonic trap are described by Legendre polynomials [34], leading to the dispersion given in Eq. (2.18) (see Sect. 2.1.2). The incommensurate spacings of mode energies inhibits any recurrence on observable time scales and lead to a complicated structure of the phase correlation dynamics [35].

This is further illustrated in Fig. 5.14b showing the evolution of the phase correlation function after a decoupling quench, calculated in the Luttinger liquid model (see Sect. 2.3.1) for the typical longitudinal trap frequency $\omega_z = 2\pi \cdot 7\,\text{Hz}$. Although partial revival features can be found, more distinct than in the measured correlations, there is no full return of the initial state. However, around $t = 210\,\text{ms}$ a signal appears whose magnitude is at least comparable to the recurrences observed in the imperfect box trap. The origin of this signal lies in the $\sqrt{j(j+1)/2}$ structure of the dispersion relation (Eq. 2.18). Higher energy modes in the limit of large j become equally spaced with $\Delta\omega = \omega_z/\sqrt{2}$, only the lowest modes are incommensurate. On a closer look however, the spectrum reveals that the modes dynamics are not completely out of sync. The main difference in energy spacings occurs for $\omega_1 = \omega_z$ and all higher modes which are nearly equally spaced with $\sim\Delta\omega$. This reduces the problem to a rephasing of these two frequencies and since $1/\sqrt{2} \approx 2/3$ they are also nearly commensurate. From the recurrence condition (Eq. 5.1)

$$\begin{aligned} \omega_1\, t_\text{rec} = \omega_z\, t_\text{rec} &= 2\pi\, n_1\ , \\ \Delta\omega\, t_\text{rec} \approx \frac{2}{3}\omega_z\, t_\text{rec} &= 2\pi\, n'\ , \end{aligned}$$

where n_1 and n' are integers denoting the number of turns the lowest mode and the number of additional turns all higher modes perform, respectively, we can see that at $t_\text{rec} = 6\pi/\omega_z$ all modes rephase close to the initial configuration. At that time the lowest lying mode performed $n_1 = 3$ turns while each higher mode performed an additional $n' = 2$ turns ($n_2 = 5$, $n_3 = 7$, $n_3 = 9, \ldots$). For a longitudinal trap frequency of $\omega_z = 2\pi \cdot 7\,\text{Hz}$ this happens after $\sim$430 ms. As in the case of the box

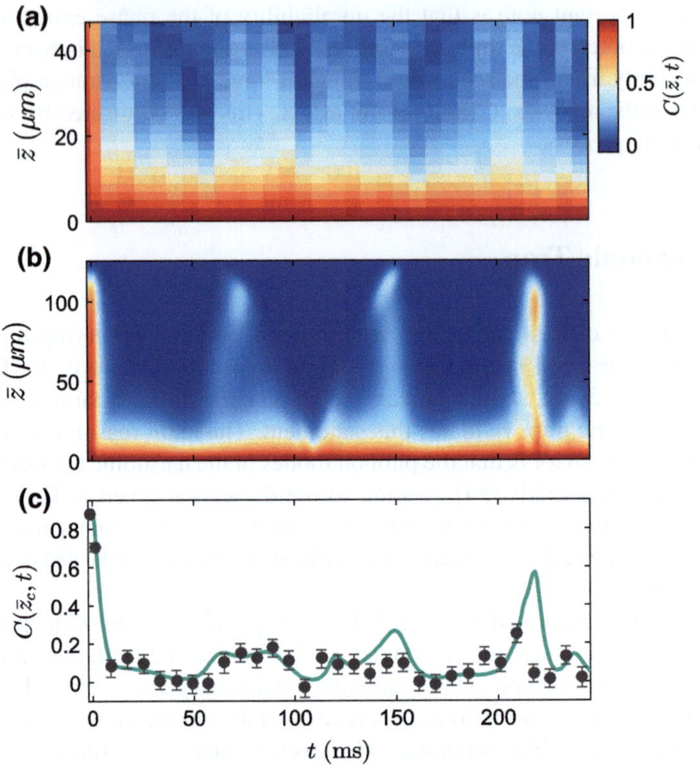

Fig. 5.14 Dephasing dynamics in the harmonic trap. a Measurement of the dephasing dynamics after decoupling in a harmonic trap with longitudinal trap frequency $\omega_z = 2\pi \cdot 7\,\text{Hz}$ and $N = 4500$ atoms in each well. Due to the inhomogeneity of the system the correlations are calculated for points located symmetrically around the center of the gas, i.e. not averaged over all points that fulfill $\bar{z} = z - z'$. **b** Luttinger liquid calculations for the same parameters as in (**a**), also showing the unaveraged two-point phase correlations. **c** Comparison of a temporal cuts of the measurement (black data points) and the Luttinger liquid calculations (green line) for $\bar{z}_c = 37\,\mu\text{m}$. The error bars of the measured data give the 68% confidence interval obtained from a bootstrap (Appendix C)

confinement, we would already expect a recurrence signal after half of that time since we start from a situation with a long range correlated relative phase field $\varphi(z)$ and the mirrored phase field $\varphi(-z)$ reached in between looks just like the initial one (see Sect. 5.1).

In the experiment, however, at that timescale phonon-phonon scattering will have destroyed most of the signal, as we have seen in Sect. 5.4, explaining why the marked feature at ~215 ms does not show up in the measurement shown in Fig. 5.14a. Nevertheless, the measured phase correlations agree stunningly well with the complex substructure before this point, as shown in Fig. 5.14c. For a tighter harmonic trap with $\omega_z > 2\pi \cdot 20\,\text{Hz}$, moving the whole dynamics to earlier times, it might therefore be feasible to observe also this reshaping feature. However, due to technical limitations of the current in the atom chips U-wires responsible for the longitudinal confinement we are limited to lower trap frequencies in our setup.

5.6 Conclusion

This chapter gave a detailed account of the observation of recurrences in a system of a few thousand particles. Beyond the fundamental interest connected to the emergence of statistical ensembles from unitary dynamics this study also opens a new window into the non-equilibrium dynamics of complex many-body systems. The recurrences provide a signal at times much past the global dephasing time. This enables the quantification of coherences at these late times and with it allows for the testing of effective models, the verification of quantum simulators or the probing of small perturbations whose effects are magnified by the long evolution times. Tracing the recurrence damping back to phonon-phonon interactions mediated by higher order terms, as done in Sect. 5.4, is a prime example for their merit as it enabled the first quantitative experimental study of phonon-phonon scattering in weakly interacting 1d Bose gases. Although the engineering of spectra that support a full recurrence at accessible time scales will not be feasible in many systems, also the realization of partial revivals gives valuable insights into the evolution of coherence and provides a promising route to investigate complex many-body systems.

On a final note, it is important to stress that the observed signal constitutes more than a revival of the involved observables. Within the low-energy description it corresponds to a return of the entire state, and therefore to a recurrence. Of course, the full many-body wave function guided by the Lieb-Lininger Hamiltonian does not return to its initial configuration on experimentally accessible time scales. However, it is also not accessible to measurements. Nevertheless, we have to remark that, although constituting a quantum Poincaré recurrence, the rephasing of modes observed in our setup is somewhat of an exceptional case. The key feature of both classical and quantum recurrence theorems is that the return inevitably happens after a long enough time, also for chaotic systems or incommensurately spaced spectra. The rephasing of the equally spaced modes in our system represents the most trivial application of these ideas.

References

1. Poincaré H (1890) Sur le problème des trois corps et les équations de la dynamique. Acta Math 13:1–270
2. Boltzmann L (1872) Weitere Studien über das Wärmegleichgewicht unter Gasmolekülen. Sitzungsberichte der Kaiserlichen Akademie der Wissenschaften Wien (II) 66:275–370
3. Zermelo E (1896) Über einen Satz der Dynamik und die mechanische Wärmetheorie. Annalen der Physik 293:485–494

4. Bocchieri P, Loinger A (1957) Quantum recurrence theorem. Phys Rev 107:337–338
5. Percival IC (1961) Almost periodicity and the quantal H theorem. J Math Phys 2:235
6. Hogg T, Huberman BA (1982) Recurrence phenomena in quantum dynamics. Phys Rev Lett 48:711–714
7. Rempe G, Walther H, Klein N (1987) Observation of quantum collapse and revival in a one-atom maser. Phys Rev Lett 58:353–356
8. Brune M et al (1996) Quantum Rabi oscillation: a direct test of field quantization in a cavity. Phys Rev Lett 76:1800–1803
9. Greiner M, Mandel O, Hänsch TW, Bloch I (2002) Collapse and revival of the matter wave field of a Bose Einstein condensate. Nature 419:51–54
10. Will S et al (2010) Time-resolved observation of coherent multi-body interactions in quantum phase revivals. Nature 465:197–201
11. Clos G, Porras D, Warring U, Schaetz T (2016) Time-resolved observation of thermalization in an isolated quantum system. Phys Rev Lett 117:170401
12. Zeiher J et al (2017) Coherent many-body spin dynamics in a long-range interacting Ising chain. Phys Rev X 7:041063
13. Schweigler T et al (2017) Experimental characterization of a quantum many-body system via higher-order correlations. Nature 545:323–326
14. Rauer B et al (2018) Recurrences in an isolated quantum many-body system. Science 360:307–310
15. Gring M et al (2012) Relaxation and prethermalization in an isolated quantum system. Science 337:1318–1322
16. Langen T, Geiger R, Kuhnert M, Rauer B, Schmiedmayer J (2013) Local emergence of thermal correlations in an isolated quantum many-body system. Nat Phys 9:640–643
17. Lewenstein M, You L (1996) Quantum phase diffusion of a Bose-Einstein condensate. Phys Rev Lett 77:3489–3493
18. Langen T et al (2015) Experimental observation of a generalized Gibbs ensemble. Science 348:207–211
19. Gogolin C, Eisert J (2016) Equilibration, thermalisation, and the emergence of statistical mechanics in closed quantum systems. Rep Prog Phys 79:056001
20. Rigol M, Dunjko V, Yurovsky V, Olshanii M (2007) Relaxation in a completely integrable many-body quantum system: an Ab Initio study of the dynamics of the highly excited states of 1D Lattice Hard-Core Bosons. Phys Rev Lett 98:050405
21. Calabrese P, Cardy J (2006) Time dependence of correlation functions following a quantum quench. Phys Rev Lett 96:136801
22. Rieger H, Iglói F (2011) Semiclassical theory for quantum quenches in finite transverse Ising chains. Phys Rev B 84:165117
23. Essler FHL, Fagotti M (2016) Quench dynamics and relaxation in isolated integrable quantum spin chains. J Stat Mech Theory Exp 2016:064002
24. Langen T, Schweigler T, Demler E, Schmiedmayer J (2018) Double light-cone dynamics establish thermal states in integrable 1D Bose gases. New J Phys 20:023034
25. Kuhnert M (2013) Thermalization and Prethermalization in an ultracold Bose Gas. PhD thesis, TU Vienna
26. Mazets IE, Schmiedmayer J (2008) Dephasing in two decoupled one-dimensional Bose-Einstein condensates and the subexponential decay of the interwell coherence. Eur Phys J B 68:335–339
27. Stimming H-P, Mauser NJ, Schmiedmayer J, Mazets IE (2011) Dephasing in coherently split quasicondensates. Phys Rev A 83:023618
28. Huber S, Buchhold M, Schmiedmayer J, Diehl S (2018) Thermalization dynamics of two correlated bosonic quantum wires after a split. Phys Rev A 97:043611
29. Grisins P, Mazets IE (2011) Thermalization in a one-dimensional integrable system. Phys Rev A 84:053635
30. Kulkarni M, Lamacraft A (2013) Finite-temperature dynamical structure factor of the one-dimensional Bose gas: From the Gross-Pitaevskii equation to the Kardar-Parisi-Zhang universality class of dynamical critical phenomena. Phys Rev A 88:021603

31. Schweigler T (2019) Correlations and dynamics of tunnel-coupled one-dimensional Bose gases. PhD thesis, TU Vienna
32. Erne S (2018) Far-from-equilibrium quantum many-body systems from universal dynamics to statistical mechanics. PhD thesis, Ruperto-Carola University of Heidelberg
33. Mazets IE, Schmiedmayer J (2010) Thermalization in a quasi-one-dimensional ultracold bosonic gas. New J Phys 12:055023
34. Petrov DS, Shlyapnikov GV, Walraven JTM (2000) Regimes of quantum degeneracy in trapped 1D gases. Phys Rev Lett 85:3745–3749
35. Geiger R, Langen T, Mazets IE, Schmiedmayer J (2014) Local relaxation and light-cone-like propagation of correlations in a trapped one-dimensional Bose gas. New J Phys 16:053034

Chapter 6
Outlook

The measurements and analysis presented in this thesis investigate two phenomena occurring in 1d Bose gases on time scales longer than the initial dephasing time: the cooling through uniform loss covered in Chap. 4 and the observation of recurrences in the relative phase fluctuations of two decoupled gases presented in Chap. 5. Both projects offer interesting routes to continue research and in the following some thoughts on possible directions will be given. Also, in Sect. 6.1, a new dipole trap setup is presented that will allow for an arbitrary and dynamic control of the longitudinal potential, opening up various possibilities for novel experiments.

For the loss induced cooling we already discussed possible routes to further investigate the absence of the shot noise contribution in Sect. 4.4. However, within the classical description promising avenues to continue research on this topic exist as well. For one, it would be interesting to experimentally investigate the non-thermal occupations arising in the high-k modes during cooling. Even though in our setup this is not feasible at the moment, a light sheet imaging capable of resolving the momentum distribution after long expansion times would make it possible [1]. Focusing the cloud in expansion through a potential lens applied by a dipole light pattern shaped accordingly could also provide access to the momentum space distribution (see Sect. 6.1). Such studies would complement the observation of non-thermal states mentioned in Refs. [2, 3] and test how well the models discussed in Chap. 4 apply to the higher energy modes.

Another interesting experiment realizable through arbitrary control over the longitudinal potential is a truly 1d evaporation where atoms can only leave the gas at the edges. Such a localized outcoupling would predominantly target high energy modes driving the system far from equilibrium or bringing non-thermal states created by homogeneous outcoupling closer to thermal equilibrium.

For the recurrence damping discussed in Chap. 5 a promising route for further investigations would be to systematically measure the recurrence height dependence on density and temperature to pin down the specific higher-order terms primarily responsible for the decay. Additional theoretical and numerical studies would be necessary to predict scaling relations between the damping rate and the samples

B. Rauer, *Non-Equilibrium Dynamics Beyond Dephasing*,
Springer Theses, https://doi.org/10.1007/978-3-030-18236-6_6

temperature and density. According to Ref. [4] the damping should become stronger for smaller densities which would be interesting to verify experimentally.

Another interesting angle would be to initialize the system in a double well with a lower tunneling coupling such that the relative phase exhibits non-Gaussian fluctuations, as discussed in Ref. [5]. With the initially Gaussian relative density fluctuations being rotated into the phase quadrature during the short time dephasing the non-Gaussianity is lost during the dynamics [6]. However, at the point of the recurrence one could expect it to return. Investigating this processes experimentally as well as characterizing the damping of the possible non-Gaussian return would be of great interest. Though, one has to note that these measurements would be challenging as the estimation of the higher-order correlations necessary would demand large statistics.

A new method to gain information through the known unitary dynamics in experiments like the ones presented in Chap. 5 is presented in Ref. [7]. In this work, a reconstruction protocol is demonstrated that allows to extract the otherwise hidden relative density fluctuations from the dynamics of the relative phase fluctuations. For the case of the recurrence measurements, this allows to quantify the full initial state of the relative degrees of freedom and to predict a recurrence from information gathered solely in between recurrences. The method opens up new measurement possibilities and might be useful in a further investigation of the recurrence damping.

6.1 Arbitary 1d Potentials

A more precise and flexible control over the confining potential in a cold atoms experiment often allows one to access completely new phenomena. For one, the realization of homogeneous samples can unveil effects otherwise hidden by an inhomogeneous density distribution [8–10], as illustrated by the results presented in Chap. 5 and Ref. [11]. Further, a local or dynamic control of the potential landscape offers new possibilities in manipulating the excitations in the system. For example, it allows for the controlled excitation of modes [12] or the realization of sonic black holes [13, 14]. In order to go beyond the possibilities of the mask setup presented in Sect. 3.2.3 we therefore designed an imaging system demagnifying the light patterns shaped by a DMD, providing us with an arbitrary control of the longitudinal trapping potential. In the following, we will shortly discuss this setup and the possibilities it provides. For a more detailed discussion see Ref. [15].

The basic principle of the applied dipole potentials remains the same as for mask setup. The objective of the transverse imaging system is used to image a shaped intensity pattern onto the atoms from a direction perpendicular to the 1d axis of the magnetic trap. Only now, the shaping of the light is performed by the array of micromirrors on the DMD chip[1] instead of the static wire mask. Each of these mirrors can be set to an *on* or *off* state, either reflecting light into the optical system that delivers it to the atoms or reflecting it out such that it is dumped. This provides a local digital control of the light intensity. To achieve arbitrary potentials, however, a smooth tuning of the local intensity is necessary. This is realized in two ways. First, the employed imaging system[2] demagnifies a single mirror down to an edge length of 0.4 μm in image space. This length is far below the diffraction limited 4 μm Airy spot diameter of the optical system, meaning that many mirrors are imaged onto a single spot in the plane of the atoms. Grayscaling can therefore be achieved by activating different numbers of mirrors within one spot. Further, each mirror can be toggled with varying on and off times. As long as the toggling frequency is far above the largest trap frequency of the transverse confinement the atoms experience an averaged intensity, providing an additional degree of freedom to tune the local potential strength.

With the DMD shaping the spatial light intensity in two dimensions while we are only interested in 1d optical potentials, additional possibilities arise to improve the grayscaling resolution. First, the DMD can be run in a special area-of-interest mode that allows to control 200 rows with a maximum pattern refresh rate of 47.6 kHz. As we only need a few tens of rows to span the transverse extension of our cloud we can make use of this mode allowing us to apply several toggling patterns whose lowest frequency lies distinctively above our maximum transverse trapping frequency of $\omega_\perp = 2\pi \cdot 2\,\text{kHz}$. Second, as we are not interested in applying transverse potentials we can implement a horizontal slit in the Fourier plane of the first 4f system to artificially degrade the resolution in the transverse direction. This leads to more pixels contribution to a single spot, improving the grayscale resolution while keeping the longitudinal spatial resolution unchanged.

The flexibility gained through the DMD setup will open a myriad of interesting paths to continue research on our 1d bosonic systems. For experiments necessitating a uniform density we will be able to counteract the harmonic magnetic potential and realize truly flat potentials over length scales much longer than the box lengths employed in Chap. 5. Further, also the magnetic potential roughness stemming from wire imperfections (see Sect. 3.2.2) can be corrected through this setup. This will allow for the creation of large systems that can be split or moved longitudinally while still retaining the flexibility of the transverse double well potential. Simple experiments could investigate how two gases with different temperature equilibrate via a longitudinal tunneling coupling and how such processes proceed locally. Also, being able to move small systems longitudinally, alternatively coupling them to dif-

[1]Texas Instruments DLP9500, 1920 × 1080 FullHD resolution with 10.8 × 10.8 μm mirror size and a maximum full pattern refresh rate of about 18 kHz. The DMD and its controllers are implemented in a V-9501 module from ViALUX.

[2]The imaging system consists of two parts. The first part is a two lens 4f systems with a 3-fold demagnification and a variable horizontal and vertical slit in the Fourier plane. The second part consists of a single lens and the objective of the transverse imaging system, forming a second 8.5-fold demagnification stage. The light is fed into transverse imaging path by a 2" polarizing beam splitter (ThorLabs PBS512).

ferent heat bathes while compressing or decompressing them should allow us to realize simple thermal machines [16]. Turning to dynamic potentials, we could further shake parts of the potential at specific frequencies to excite only certain modes in the system. This would allow us to perform phonon spectroscopies and to monitor the modes decay individually, providing a much cleaner measurement of the damping effects found in Fig. 5.10 of Chap. 5. As the optical potential is independent of the magnetic confinement it can also be used to manipulate the atoms in their initial expansion phase after the trap has been turned off. This should enable the implementation of optical potential lenses that map the atomic momentum distribution into real space during the expansion. Finally, as already mentioned above, the shaping of condensate flow through specially designed dynamic potentials opens the possibility to investigate sonic black hole analogues and the associated effects of phononic Hawking radiation [17, 18].

References

1. Bücker R et al (2009) Single-particle-sensitive imaging of freely propagating ultracold atoms. New J Phys 11:103039
2. Fang B, Carleo G, Johnson A, Bouchoule I (2014) Quench-induced breathing mode of one-dimensional Bose gases. Phys Rev Lett 113:035301
3. Fang B, Johnson A, Roscilde T, Bouchoule I (2016) Momentum-space correlations of a one-dimensional Bose gas. Phys Rev Lett 116:050402
4. Stimming H-P, Mauser NJ, Schmiedmayer J, Mazets IE (2011) Dephasing in coherently split quasicondensates. Phys Rev A 83:023618
5. Schweigler T et al (2017) Experimental characterization of a quantum many-body system via higher-order correlations. Nature 545:323–326
6. Schweigler T, et al (Unpublished) Gaussification in a cold atomic quantum simulator
7. Gluza M, et al (2018) Quantum read-out for cold atomic quantum simulators. arXiv:1807.04567
8. Gaunt AL, Schmidutz TF, Gotlibovych I, Smith RP, Hadzibabic Z (2013) Bose-Einstein Condensation of Atoms in a Uniform Potential. Phys Rev Lett 110:200406
9. Mukherjee B et al (2017) Homogeneous atomic Fermi gases. Phys Rev Lett 118:123401
10. Hueck K et al (2018) Two-dimensional homogeneous Fermi gases. Phys Rev Lett 120:60402
11. Rauer B et al (2018) Recurrences in an isolated quantum many-body system. Science 360:307–310
12. Ville JL et al (2018) Sound propagation in a uniform superfluid two-dimensional Bose gas. Phys Rev Lett 121:145301
13. Unruh WG (1981) Experimental black-hole evaporation? Phys Rev Lett 46:1351–1353
14. Lahav O et al (2010) Realization of a sonic black hole analog in a Bose-Einstein condensate. Phys Rev Lett 105:240401
15. Tajik M (2017) Arbitrary one-dimensional optical dipole potentials on an atom chip. Master's thesis, TU Vienna
16. Kosloff R, Levy A (2014) Quantum heat engines and refrigerators: continuous devices. Annu Rev Phys Chem 65:365–393
17. Steinhauer J (2014) Observation of self-amplifying Hawking radiation in an analogue black-hole laser. Nat Phys 10:864–869
18. Steinhauer J (2016) Observation of quantum Hawking radiation and its entanglement in an analogue black hole. Nat Phys 12:959–965

Appendix A
Transverse Broadening Corrections

In Sect. 2.2.2 we discussed the corrections due to transverse interactions derived from a variational approach [1]. Here, we will show the full perturbative expansion for a single gas going up to third order. The calculation closely follows the approach of Mora and Castin [2] and takes the discretization necessary for the definition of the phase operator into account.

From the expression for the width of the transverse wave function we can calculate the mean field interaction energy term for the effective longitudinal Hamiltonian to be

$$E_{\text{int}} = \hbar\omega_\perp\sqrt{1 + 2a_s n_{1\text{d}}}\ .$$

Requantizing this result and inserting it in the a discretized form of the Hamiltonian introduced in [3] we obtain

$$\hat{H} = \sum_z l\ \hat{\psi}_z^\dagger\left[-\frac{\hbar^2}{2m}\Delta_z + U_z - \mu + \hbar\omega_\perp\sqrt{1 + 2a_s\hat{\psi}_z^\dagger\hat{\psi}_z}\right]\hat{\psi}_z\ .$$

Here, z is the index of position on the grid and l is the dicretization length. We insert a phase-density representation of the field operator on each grid point $\hat{\psi}_z = e^{i\hat{\theta}_z}\sqrt{\hat{n}_z} = e^{i\hat{\theta}_z}\sqrt{n_{1\text{d}} + \delta\hat{n}_z}$ and perturbativly expand the Hamiltonian around small density and long range phase fluctuations. The spatial indices of the fields will be omitted from now on for clarity.

Our interest lies in the modifications the broadening causes in the potential energy terms. Inserting the phase-density representation we obtain

$$\begin{aligned}
\hat{H}_p &= \sum_z l\left[(U-\mu)\hat{\psi}^\dagger\hat{\psi} + \hbar\omega_\perp\hat{\psi}^\dagger\sqrt{1 + 2a_s\hat{\psi}^\dagger\hat{\psi}}\hat{\psi}\right] \\
&= \sum_z l\ \hat{n}\left[(U-\mu) + \hbar\omega_\perp\sqrt{1 + 2a_s\left(\hat{n} - \frac{1}{l}\right)}\right]. \qquad \text{(A.1)}
\end{aligned}$$

B. Rauer, *Non-Equilibrium Dynamics Beyond Dephasing*,
Springer Theses, https://doi.org/10.1007/978-3-030-18236-6

Here we used the identity $\hat{\psi}^\dagger\hat{\psi} = \hat{n}$ and the commutator of the field operators $[\hat{\psi}_z, \hat{\psi}^\dagger_{z'}] = \frac{\delta_{z,z'}}{l}$. Expanding the square root in this expression for small $a_s\hat{n}$ we can rewrite Eq. (A.1) as

$$\hat{H}_p = \sum_z l\,(n_{1d} + \delta\hat{n})\Big[(U-\mu) + \hbar\omega_\perp\Big(1 + a_s\Big(n_{1d} + \delta\hat{n} - \frac{1}{l}\Big) - \frac{1}{2}a_s^2\Big(n_{1d} + \delta\hat{n} - \frac{1}{l}\Big)^2 + \dots\Big)\Big].$$

Collecting all terms of order $O(\delta\hat{n}^n)$ with $n \leq 3$, considering that $1/l$ is of second order and associating the obtained series with closed expressions we obtain[1]

$$\begin{aligned}
\hat{H}_p^{(0)} &= \sum_z l\, n_{1d}\Big[(U-\mu) + \hbar\omega_\perp\sqrt{1+2a_s n_{1d}}\Big]\\
\hat{H}_p^{(1)} &= \sum_z l\, \delta\hat{n}\Big[(U-\mu) + \hbar\omega_\perp \frac{1+3a_s n_{1d}}{\sqrt{1+2a_s n_{1d}}}\Big]\\
\hat{H}_p^{(2)} &= \sum_z l\, \delta\hat{n}^2\, \frac{\hbar\omega_\perp a_s}{2}\Big[\frac{2+3a_s n_{1d}}{(1+2a_s n_{1d})^{3/2}}\Big] - \sum_z \frac{\hbar\omega_\perp a_s n_{1d}}{\sqrt{1+2a_s n_{1d}}}\\
\hat{H}_p^{(3)} &= -\sum_z l\, \delta\hat{n}^3\, \frac{\hbar\omega_\perp a_s^2}{2}\Big[\frac{2a_s^2 n_{1d}^2 - a_s n_{1d} + 1}{(1+2a_s^2 n_{1d}^2)^{5/4}}\Big] -\\
&\quad - \sum_t \delta\hat{n}\, \hbar\omega_\perp a_s\Big[\frac{2a_s^2 n_{1d}^2 - a_s n_{1d} + 1}{(1+2a_s^2 n_{1d}^2)^{3/2}}\Big].
\end{aligned} \tag{A.2}$$

In zeroth order we regain the broadened potential energy for the average density profile n_{1d}. In first order we obtain the broadened expression for the chemical potential given in Eq. (2.33). Therefore, as in the unbroadened case, $H^{(1)}$ vanishes for the density profile that minimizes $H^{(0)}$, i.e. solves the non-polynomial GPE in Eq. (2.34). This again leaves the leading terms governing the fluctuations to be of second order. They are of the same form as in the purely 1d case but their prefactors exhibit a more complicated density dependence. The first term in $H_p^{(2)}$ reproduces g_n defined in Eq. (2.35) and thereby confirms the expression for the speed of sound found from the hydrodynamics relation.

In third order, apart from the rescaled prefactor of the last term in Eq. (A.2) which is also appears in the purely 1d case, the first new term arises. It is proportional to $\delta\hat{n}^3$ and represents the lowest order integrability breaking that arises from the broadening. Combining this term with the second order term, neglecting the last term in Eq. (A.2) and disregarding the constant energy functional in $H_p^{(2)}$ we end up with

$$\hat{H}_p^{(2+3)} = \sum_z l\, \frac{g_n}{2}\delta\hat{n}^2\left(1 - \chi_n a_s \delta\hat{n}\right),$$

[1] The closed expressions were obtained through the Mathematica function 'FindGeneratingFunction'.

where $\chi_n \approx 1/2$ for $a_s n_{1d} < 1$. As $a_s \delta\hat{n}$ is small, this correction will not have large effects on equilibrium states. For long time dynamics like the loss of coherence observed in the measurements presented in Chap. 5 in might play a larger role. However, as discussed in Sect. 2.2.3, a more rigorous treatment of the integrability breaking effects stemming from transverse interactions performed in Refs. [3, 4] concluded that also the effects on intermediate time dynamics are negligible for our parameter range.

References

1. Salasnich L, Parola A, Reatto L (2002) Effective wave equations for the dynamics of cigar-shaped and disk-shaped Bose condensates. Phys Rev A 65:043614
2. Mora C, Castin Y (2003) Extension of Bogoliubov theory to quasicondensates. Phys Rev A 67:053615
3. Mazets IE, Schumm T, Schmiedmayer J (2008) Breakdown of integrability in a quasi-1D ultra-cold bosonic gas. Phys Rev Lett 100:210403
4. Mazets IE, Schmiedmayer J (2010) Thermalization in a quasi-one-dimensional ultracold bosonic gas. New J Phys 12:055023

Appendix B
Radial GPE Simulations

Solving the GPE numerically in more than one dimension can become computationally demanding especially when interested in dynamical effects. However, in certain cases the exploitation of symmetries can cast the problem into a lower dimensional form. In this appendix, we will discuss how the 2d GPE can be solved numerically for a radially symmetric problem with the aid of Fourier-Bessel series. The approach maps the 2d problem onto a 1d radial problem thereby significantly reducing the its computational demands.

For an arbitrary potential $U(r)$ the radially symmetric 2d GPE can be written as

$$i\hbar\frac{\partial}{\partial t}\phi(r,t) = -\frac{\hbar^2}{2m}\Delta_r\,\phi(r,t) + U(r)\,\phi(r,t) + g|\phi(r,t)|^2\,\phi(r,t)\;, \tag{B.1}$$

with $r^2 = x^2 + y^2$ and $\Delta_r = \frac{1}{r}\frac{\partial}{\partial r}r\frac{\partial}{\partial r}$ being the radial part of the Laplace operator in polar coordinates. A common strategy to numerically solve the GPE is the split-step method [1]. There, the wave function is propagated alternatingly by the kinetic term of the Hamiltonian and the terms diagonal in real space, using small time steps to minimize the error. The propagation with the kinetic term is usually performed in momentum space where it becomes diagonal. This strategy can also be applied to solve Eq. (B.1), only that instead of obtaining the momentum space wave function through a Fourier transform one uses the Fourier-Bessel series [2], expanding $\phi(r)$ in Bessel functions of the first kind J_α

$$\phi(r) = \sum_{j=1}^{\infty}\psi_j\,J_0(\beta_j\,r)\;, \tag{B.2}$$

$$\psi_j = \frac{2}{\left[L\,J_1(u_{0,j})\right]^2}\int_0^L r\,\mathrm{d}r\;\phi(r)\,J_0(\beta_j\,r)\;. \tag{B.3}$$

Here, $\beta_j = \frac{u_{0,j}}{L}$ are the radial momenta with $u_{0,j}$ being the j-th root of $J_0(r)$ and L the radial size of the system. The pre-factor in front of the integral in Eq. (B.3) is a

B. Rauer, *Non-Equilibrium Dynamics Beyond Dephasing*,
Springer Theses, https://doi.org/10.1007/978-3-030-18236-6

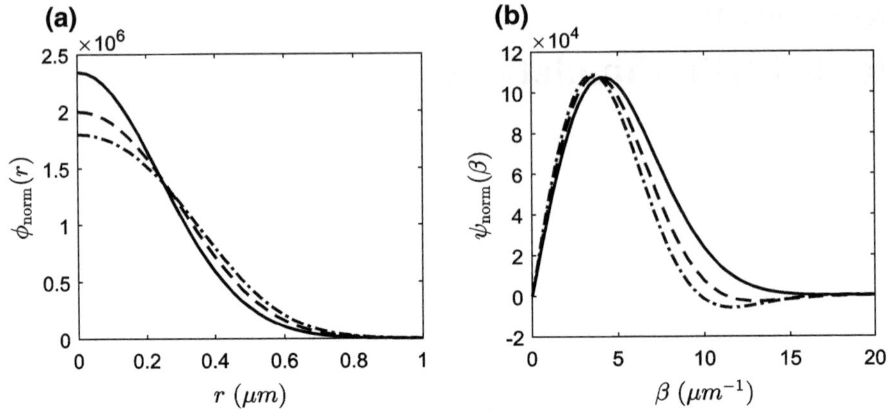

Fig. B.1 **Radial wave function.** **a** Normalized wave function $\phi_{\rm norm}(r) = \phi(r)/\sqrt{\int |\phi(r)|^2 r \, {\rm d}r}$ of ^{87}Rb atoms trapped in a harmonic confinement with $\omega_\perp = 2\pi \cdot 2\,$kHz for three different longitudinal densities $n_{1{\rm d}} a_{\rm s} = 0$, 0.3 and 0.6 (solid, dashed and dash-dotted line). **b** Corresponding Fourier-Bessel coefficients representing the radial momentum distribution

normalization stemming from

$$\int_0^L r \, {\rm d}r \; J_0(\beta_i \, r) \; J_0(\beta_j \, r) = \delta_{ij} \frac{\left[L \; J_1(u_{0,j})\right]^2}{2} \, .$$

Using the fact that $J_0(r)$ is a solution of Bessel's differential equation it can be shown that

$$\Delta_r \, \phi(r) = \sum_{j=1}^{\infty} \psi_j \; \Delta_r \; J_0(\beta_j \, r) = - \sum_{j=1}^{\infty} \psi_j \; \beta_j^2 \; J_0(\beta_j \, r) \, ,$$

rendering the kinetic term in Eq. (B.1) diagonal for the momentum space wave function ψ_j. This allows for an effective implementation of the kinetic propagation through the transformations defined in Eqs. (B.2) and (B.3).

As for Cartesian coordinates, the ground state of the system can be found through imaginary time evolution. Figure B.1 shows ground states for different numbers of particles in an harmonic potential. Dynamical solutions of Eq. (B.1) are discussed in the context of the transverse expansion of clouds released from the tight atom chip confinement (see Sect. 3.3.1). It is important to note that this approach is limited to fully radially symmetric problems. For example, dipole oscillations of a condensate trapped in a otherwise radially symmetric potential can not be simulated.

This approach can further be extended to cylindrical coordinates (r, z) to effectively solve 3d problems. There, a combination of Fourier transform and Fourier-Bessel series can be used to obtain the momentum space wave function.

References

1. Bao W, Jaksch D, Markowich PA (2003) Numerical solution of the Gross-Pitaevskii equation for Bose-Einstein condensation. J Comput Phys 187:318–342
2. Kaplan W (2002) Advanced calculus, 5th edn. Pearson

References

1. Bao W, Jaksch D, Markowich PA (2003) Numerical solution of the Gross-Pitaevskii equation for Bose-Einstein condensation. J Comput Phys 187:318–342
2. Kaplan W (2002) Advanced calculus, 5th edn. Pearson

Appendix C
Bootstrap

Bootstrapping is a resampling technique that allows in many cases to assert the accuracy of a parameter estimation even for complex estimators where an accuracy measure (i.e. confidence interval, bias,...) is hard to calculate analytically [1, 2]. Its core premise is to assume the empirical distribution of measured data to be a good approximation of the unknown true distribution underlying the data. From this empirical distribution the statistical quantities are then calculated, generally by resampling the available data with replacement. In the following, first an intuitive example is given before a typical use case from the data analysis presented in this thesis is described.

Let us consider a set of measured data $\{x_1, \ldots, x_n\}$ from which we infer the sample mean, $\bar{x} = \frac{1}{n}\sum_i x_i$. If we want to know how accurate the value is we obtained we can perform a bootstrap. For that, we resample the data with replacement and thereby generate a set of bootstrap samples each containing again n values. Looking at one of these bootstrap samples, some of the original values will most likely appear multiple times and some will not appear at all. Also, the probability that the original sample is reproduced is very small if n is large. From each of these bootstrap samples we can now infer its sample mean, which gives us the bootstrap distribution. This distribution gives a measure of accuracy for the value $\bar{x}$ estimated from the original sample and allows us to obtain confidence intervals. In this simple example one could have also directly calculated the standard error of the mean $\bar{\sigma} = \sqrt{\sum_i (x_i - \bar{x})^2 / n(n-1)}$. However, if the estimator of interest is more complicated and such an expression is not known the bootstrap provides a simple technique to infer measures of accuracy.

In this thesis, bootstrapping is used to obtain confidence intervals for parameter estimations which involve correlation functions or fits to features of correlation functions. An example for measured data that is resampled are sets of phase profiles $\{\varphi_1(z), \ldots, \varphi_n(z)\}$, where each spatial profile $\varphi_i(z) = [\varphi_i(z_1), \varphi_i(z_2), \ldots, \varphi_i(z_m)]$ is a vector extracted from a single interference picture (see Sect. 3.3.4). Here, m is the number of pixels for which the phase is extracted and z_j are the positions in the cloud corresponding to the individual pixels. Typically, like in the recurrence measurements presented in Chap. 5, multiple of these phase profile sets are taken at different evolution times t. Let us now consider the case of the recurrence height

B. Rauer, *Non-Equilibrium Dynamics Beyond Dephasing*,
Springer Theses, https://doi.org/10.1007/978-3-030-18236-6

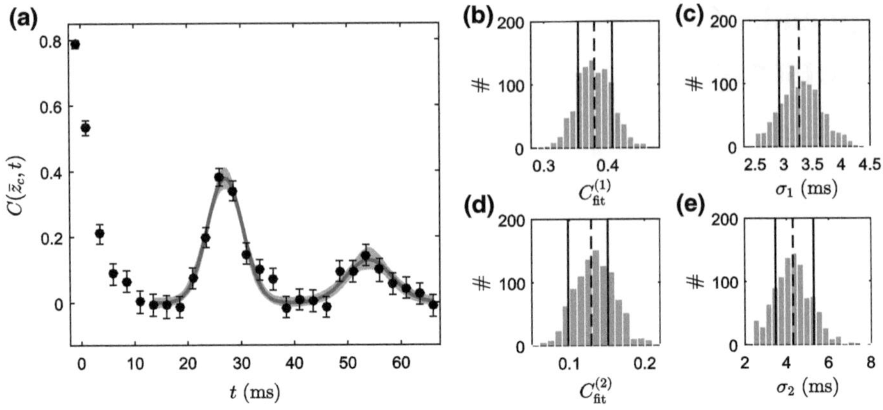

Fig. C.1 Bootstrap example. a Fit of the recurrence height for a system with $T_{\text{eff}} = 72\,\text{nK}$ and $L = 49\,\mu\text{m}$, presented earlier in Fig. 5.10c. The shaded area shows the fit values between the 10th and the 90th percentile of the bootstrap distribution for 999 runs. The fit function is given in Eq. (5.5). **b, d** Corresponding distributions of fitted recurrence heights $C_{\text{fit}}^{(1,2)}$ of the first and second recurrence, respectively. The dashed lines indicate the values estimated from the original sample while the solid lines give the obtained 68% confidence intervals. **c, e** Same for the recurrence widths $\sigma_{1,2}$

fits presented in Fig. 5.10. There we want to fit a feature in the dynamics of the phase correlation function $C(\bar{z}, t)$ which is calculated from the phase profiles (see Eqs. (2.25) and (5.2)). For that, we draw samples from the full measurement runs, where one run includes one profile from each evolution time.[2] From these samples we calculate the phase correlations and fit the double Gaussian function given in Eq. (5.5) to the spatial cut $C(\bar{z}_c, t)$. Doing this for many resampled configurations, typically around 1000, we obtain a distribution of the fitted recurrence heights $C_{\text{fit}}^{(1,2)}$ and recurrence width $\sigma_{1,2}$ (Fig. C.1). From these distributions we can infer confidence intervals which translate, for example, to the confidence intervals of H_{rec} shown in Fig. 5.13. This is done by the bias-corrected and accelerated method [1] implemented by the Matlab function 'bootci'.

References

1. Efron B, Tibshirani R (1986) Bootstrap methods for standard errors, confidence intervals, and other measures of statistical accuracy. J Stat Sci 1:54–75
2. Bohm G, Zech G (2010) Introduction to statistics and data analysis for physicists. Verlag Deutsches Elektronen-Synchrotron

[2]These profiles are independent of each other and therefore could, in principle, also be sampled independently. However, in the current implementation of Matlabs 'bootci' it is not possible to sample from multiple sets.

MIX
Papier aus verantwortungsvollen Quellen
Paper from responsible sources
FSC® C105338

If you have any concerns about our products,
you can contact us on
ProductSafety@springernature.com

In case Publisher is established outside the EU,
the EU authorized representative is:
Springer Nature Customer Service Center GmbH
Europaplatz 3, 69115 Heidelberg, Germany

Printed by Libri Plureos GmbH
in Hamburg, Germany